COLLEGE OF AEROSPACE DOCTRINE,
RESEARCH AND EDUCATION

AIR UNIVERSITY

Flying Reactors
The Political Feasibility of Nuclear Power in Space

JAMES R. DOWNEY
Lieutenant Colonel, USAFR

ANTHONY M. FORESTIER
Wing Commander, RAAF

DAVID E. MILLER
Lieutenant Colonel, USAF

CADRE Paper No. 22

Air University Press
Maxwell Air Force Base, Alabama 36112-6615

April 2005

Air University Library Cataloging Data

Downey, James R.
    Flying reactors : the political feasibility of nuclear power in space / James R. Downey, Anthony M. Forestier, David E. Miller.
        p. : cm. – (CADRE paper, 1537-3371 ; 22)
    Includes bibliographical references.

    1. Space vehicles—Nuclear power plants. 2. Nuclear power—Public opinion. 3. Nuclear power—Government policy. I. Title. II. Forestier, Anthony M. III. Miller, David E., 1960– IV. Series. V. Air University (US). College of Aerospace Doctrine, Research and Education.

629.4753—dc22

Disclaimer

Opinions, conclusions, and recommendations expressed or implied within are solely those of the authors and do not necessarily represent the views of Air University, the United States Air Force, the Department of Defense, or any other US government agency. Cleared for public release: distribution unlimited.

This CADRE Paper and others in the series are available electronically at the Air University Research Web site http://research.maxwell.af.mil and the AU Press Web site http://aupress.maxwell.af.mil.

CADRE Papers

CADRE Papers are occasional publications sponsored by the Airpower Research Institute of Air University's College of Aerospace Doctrine, Research and Education (CADRE). Dedicated to promoting the understanding of air and space power theory and application, these studies are published by Air University Press and broadly distributed to the US Air Force, the Department of Defense and other governmental organizations, leading scholars, selected institutions of higher learning, public-policy institutes, and the media.

All military members and civilian employees assigned to Air University are invited to contribute unclassified manuscripts that deal with air and/or space power history, theory, doctrine or strategy, or with joint or combined service matters bearing on the application of air and/or space power.

Authors should submit three copies of a double-spaced, typed manuscript and an electronic version of the manuscript on removable media along with a brief (200-word maximum) abstract. The electronic file should be compatible with Microsoft Windows and Microsoft Word—Air University Press uses Word as its standard word-processing program.

Please send inquiries or comments to
Chief of Research
Airpower Research Institute
CADRE
401 Chennault Circle
Maxwell AFB AL 36112-6428
Tel: (334) 953-5508
DSN 493-5508
Fax: (334) 953-6739
DSN 493-6739
E-mail: cadre.research@maxwell.af.mil

Contents

Chapter		Page
	DISCLAIMER	ii
	FOREWORD	vii
	ABOUT THE AUTHORS	ix
	ACKNOWLEDGMENTS	xi
1	WHITHER SPACE NUCLEAR POWER?	1
	Notes	11
2	SPACE NUCLEAR POWER AS TRANSSCIENTIFIC PUBLIC POLICY	13
	Notes	19
3	POLITICAL PERMISSION—THE CONTEMPORARY DIMENSIONS	21
	Notes	43
4	A TRANSSCIENTIFIC POLITICAL ENGAGEMENT STRATEGY	47
	Notes	67
5	CONCLUSIONS	69
	Notes	74

Appendix		
A	THE HISTORY OF SPACE NUCLEAR POWER	75
	Notes	87
B	PROJECT PROMETHEUS—FREQUENTLY ASKED QUESTIONS—DECEMBER 2003	89
C	THE MEMBER GROUPS OF THE FLORIDA COALITION FOR PEACE AND JUSTICE OR STOP CASSINI! CAMPAIGN	103
	SELECTED BIBLIOGRAPHY	107

Illustrations

Figure		Page
1	Required power level versus mission duration for space applications	4
2	Space nuclear power program feasibility	9
3	Public attitudes toward selected technologies, 2002	23
4	Public assessment of space exploration, 1985 to 2001	25
5	Societal level of interest in SNP: Illustrating neutral bias	28
6	Societal bias and activation sought by program proponents	29
7	Societal bias and activation sought by program opponents	29
8	Citizens who trust the United States federal government, 1958 to 2002	33
9	SNP program feasibility space	49

Tables

1	Key Characteristics of Solar, RTG, and Fission Reactor Space Power Systems	5
2	Traits of Democracies	41
3	Potential Stakeholders in a Values-Focused Decision Strategy	56

Foreword

One of the challenges Gen John P. Jumper, chief of staff of the Air Force, sends to Air Force students, researchers, and staff offices is to investigate future concepts of operations (CONOPS). One in particular relates to this study, the CONOPS for space and command, control, communications, computers, intelligence, surveillance, and reconnaissance. The Air Force is very sensitive about incorporating new technology into its operations. While the authors advocate a feasibility study for reactors in space in a CONOPS, they also explore a deeper problem with widespread societal rejection and revulsion concerning the theoretical employment of nuclear technology in space.

They point first to the mission enabling advantages of nuclear reactors in space—factors like light weight, high power, long life, and potentially lower costs. A reactor would supply electrical power to a space vehicle and perhaps provide ionic or electrical propulsion. They see that nuclear-powered spacecraft would serve long-range National Aeronautics and Space Administration (NASA) missions as well as permit effective hyperspectral satellites that would have profound benefits for the Department of Defense.

The limiting factors for nuclear power in space are a compelling mission requirement and broad acceptance in popular support. The first factor is rather obvious but the second is driven by a broad-based fear of risks in the employment of nuclear technology. Many have general doubts about such an undertaking. Some opponents perceive cataclysmic dangers. A failure of a space launch carrying nuclear systems would produce something on the order of a "dirty" nuclear bomb. Opponents are rigorous in their protest. Two things were clear to these researchers. One, nuclear space developers must convince the public that they are capable of developing a safe and robust system.

Two, because the political battle is primarily over perceived risks rather than empirically based understanding, employment of a values-focused decision strategy is necessary to convince the public and congressional leaders of the feasibility of a space nuclear program.

Flying Reactors: The Political Feasibility of Nuclear Power in Space was written as part of the Air Force Fellows research requirement. The College of Aerospace Doctrine, Research and Education (CADRE) is pleased to publish this study as a CADRE Paper and thereby make it available to a wider audience within the Air Force and beyond.

DANIEL R. MORTENSEN
Chief of Research
Airpower Research Institute, CADRE

About the Authors

Lt Col James R. Downey, PhD, is a member of the Air Force Reserve and is assigned to the Air War College Center for Strategy and Technology at Maxwell AFB, Alabama. In his civilian career, Dr. Downey is professor of Science and Technology at the Army War College, Carlisle Barracks, Pennsylvania. During fiscal year 2004, Colonel Downey was a National Security Fellow at the JFK School of Government at Harvard University. Prior to the start of his Air Force Fellowship, Colonel Downey was chairman of the Physics Department, Grove City College, Grove City, Pennsylvania. His previous AF Reserve assignment was as an adjunct faculty member, Air Command and Staff College, Maxwell AFB. He has a broad background in national security policy, nuclear physics, systems analysis, and managing scientific research and development programs. Colonel Downey holds a graduate degree in nuclear engineering and a doctorate in aerospace engineering. He is a graduate of Squadron Officer School and Air Command and Staff College, as well as the AF Fellowship Program.

Wing Cdr Anthony M. Forestier, Medal of the Order of Australia, is a Royal Australian Air Force (RAAF) flight-test navigator currently assigned to Australia's Defence Strategy Group as deputy director of Military Strategy. He was a National Security Fellow at Harvard University, class of 2004, and before that commanded the School of Air Navigation. He has been a faculty member at the RAAF Command and Staff College, and was a chief of Air Force Fellows in 1990 where he published a monograph *Into the Fourth Dimension: A Guide to Space*. Commander Forestier has operational experience on C-130 Hercules and P-3C Orion aircraft, flight test experience on a variety of aircraft including Blackhawk and F-111C, and is a flight navigator and postgraduate weapons/avionics systems instructor. He is an experienced staff officer with a background in strategy development and implementation, force development, defense science, and training. He holds graduate degrees in defense studies, management, and aerosystems. Commander Forestier is a graduate of Harvard University; the University of

Canberra; and the Royal Air Force College, Cranwell, England. He was awarded an Order of Australia medal in 2004.

Lt Col David E. Miller is currently assigned to the US European Command Strategy, Plans, and Programs Directorate. During fiscal year 2004, Colonel Miller was a National Security Fellow at the JFK School of Government at Harvard University. Prior to the start of his fellowship, Colonel Miller was commander of the 1st Reconnaissance Squadron at Beale AFB, California. He has also served as the dual-hatted commander of the 99th Expeditionary Reconnaissance Squadron and Detachment 1 of the 9th Operations Group during Operation Enduring Freedom and Operation Iraqi Freedom. He has a broad background in combat operations, national security policy formulation, military campaign strategy, systems analysis, experimental flight test, and managing military acquisition programs. Colonel Miller holds graduate degrees in systems management and airpower studies. He is a graduate of Squadron Officer School and Air Command and Staff College. He is also a graduate of the School of Advanced Airpower Studies and the USAF Test Pilot School.

Acknowledgments

The authors wish to express their sincere appreciation to Ms. Mary Schumacher and Lt Col Courtney Holmberg for their assistance in the preparation of this manuscript. The authors also wish to thank the many people who agreed to be interviewed in support of this project: Victoria Friedensen and Col Karl Walz of NASA, Mr. Bob Shaw of the Office of the Secretary of Defense, Dr. Roald Segdeev of the University of Maryland, Dr. Karl Mueller of RAND, Dr. Nicholas Miclovitch of Air Force XP (Pentagon Office of Strategic Plans), and Dr. George Ullrich of Science Applications International Corporation. Finally, the authors are indebted to their families for their support during the many hours of research, writing, and editing that were necessary to complete this project.

Chapter 1

Whither Space Nuclear Power?

Our ordinary citizens, though occupied with the pursuits of industry, are still fair judges of public matters; . . . and instead of looking on discussion as a stumbling block in the way of action, we think of it as an indispensable preliminary to any wise action at all.

—Pericles

This paper addresses the question: What mechanism(s) would improve the political feasibility of a nuclear power program for US space operations? For a period of more than 50 years, the United States has been exploring the potential of nuclear-powered reactors for use in a variety of space-based applications. From the earliest days, there have been numerous challenges—some technical, many political—that have impeded progress in every program that the United States has considered. The issues surrounding space nuclear power (SNP) are complex and multifaceted. For the United States, the development of SNP lies at the intersection of program cost benefit and the social perception of risk. The actual decision to employ SNP is finally political—encompassing political judgment, will, and acceptance of risk. But if the current climate surrounding nuclear use remains manifest, the future for SNP looks politically challenging.

The specter of a Delta IV rocket carrying a nuclear-powered satellite exploding on launch from Florida is an outcome the US government and its agencies would rather not confront. Though that has never happened, it remains the type of image that the anti-SNP lobby, under the umbrella of groups like the Florida Coalition for Peace and Justice (FCPJ), presents as a potential outcome of SNP programs.

The FCPJ cites past space-based nuclear incidents and a lack of public confidence in government agencies to combine nuclear and space technologies safely as a cause for serious public concern. Their premise is that the reward promised by programs needing or wanting to take advantage of the operational benefits offered by SNP does not outweigh the risk of adverse environmental outcomes. They believe that, as a corollary to SNP for space science, the United States is committed to weaponizing space. The FCPJ and its allies do not trust the government, and they seek to align others to their cause to stop all SNP programs, building on the disquiet felt by many US citizens about nuclear power. The FCPJ advances its cause via political and legal arenas, specifically by engaging in public protest, political activism, lobbying, and legal challenge.

On the other side of the debate, the US National Aeronautics and Space Administration (NASA) would like to fire the imagination of the public with the glories of space exploration to ensure the longevity of the space program and to counter the public's disquiet to enable missions requiring SNP. Fascinating images of Mars, Saturn, Jupiter, Neptune, and Uranus fill TV screens and adorn the front covers of glossy magazines. People are eager for information about space. In 2012 NASA's proposed SNP-based Jupiter Icy Moon Orbiter (JIMO) may use advanced sensors to search under the icy surface of Jupiter's moons Callisto, Ganymede, and Europa.[1] In the water believed to be under the ice, there may be an answer to the question: Is there life on other worlds? Space science, enabled by SNP, may soon address one of our most profound questions.

From the perspective of the Department of Defense (DOD) perhaps in the decade 2010–2020, the United States and its allies may take comfort in the fact that—although terrorism has not been eliminated—a constellation of large, long-lived SNP satellites with their hyperspectral sensors have made the problem of global intelligence, surveillance, and reconnaissance (ISR) more manageable. Terrorists and proliferators of nuclear weapons and associated delivery systems will find fewer opportunities to act and places to hide.

Is SNP an environmental menace or a feasible enabling technology? The argument is polarized in the United States, the epicenter of the debate as the world's most capable spacefaring and democratic nation. There are valid arguments for both sides. Each side of the debate has its active proponents, supported by allies and ad hoc coalitions of stakeholders. Yet, between the interlocutors in the debate there is the vast, unaligned, and politically passive or inactive majority. The public is interested in space science but is also sensitive to the costs and risks. Politically aligned and activated, even a small part of that majority would pose pressure that policy makers in the government could not ignore, and such pressure may determine the feasibility of SNP systems going forward.

Despite the polarization in the public debate about SNP, there is no doubt about the attractiveness of the technology to support space-based missions. Space science and national security are both missions enabled by the next generation of satellites and space vehicles. Such vehicles may depend on onboard nuclear reactors to reliably generate large amounts of electricity for power and propulsion.

Motivation for Space Nuclear Power

All space vehicles require onboard power sources. For most space systems, a combination of batteries and solar panels provides onboard electrical power. Figure 1 illustrates the characteristics that alternative energy sources offer for space missions.

A few examples of how this graph is interpreted should prove useful in explaining its meaning. If a particular mission requires 1,000 kilowatts of electric power (kWe) and must last for one day, then chemical batteries are suitable for accomplishing this mission. A system such as the international space station, which requires approximately 240 kWe and must function for several years, uses solar panels.[2] The graph shows that for space missions demanding both high power (100 kWe and more) and long duration (months to years), fission reactors offer the only existing practicable option for providing electrical energy to the spacecraft. Reactors also promise power for thrust. These capabilities are attractive and make SNP desirable as an enabling technology for multiple purposes. Radioisotope thermoelectric generators (RTG), a commonly used form of SNP, offer less power than reactors but an equally long life, and they have utility for deep space missions where the Sun's energy density is too low for solar panels to be

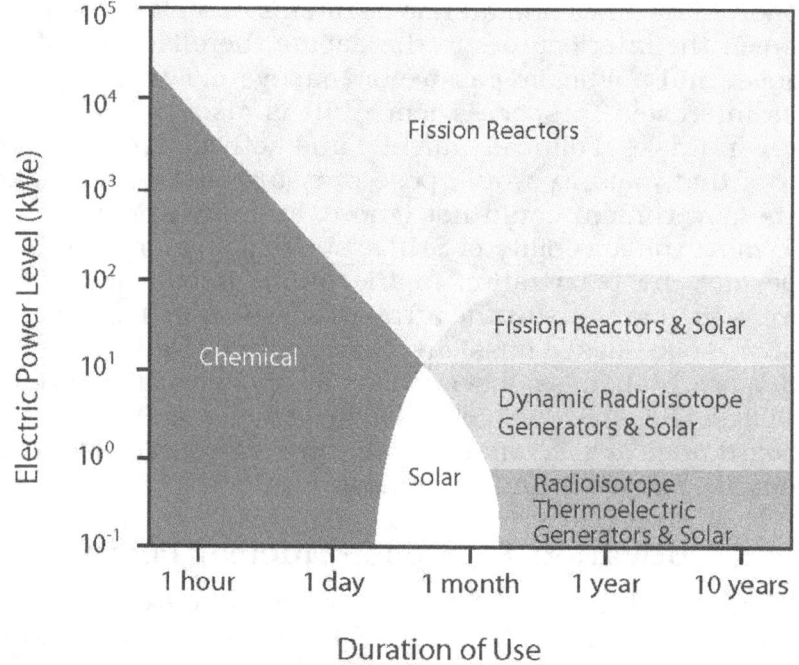

Figure 1. Required power level versus mission duration for space applications. (Reprinted from Joseph A. Angelo and David Buden, *Space Nuclear Power* [Malabar, Fla.: Orbit Book Company, 1985], ix.)

effective. Table 1 summarizes the key technical characteristics of SNP compared to solar power.

SNP's technological advantages make it attractive as a power source. These significant advantages raise the question as to why the United States currently does not have any SNP systems. History provides insight to the contemporary argument.

Space Nuclear Power History and Its Implications

The United States has had a public policy interest in the development of SNP since the late 1950s. Over the years there have been several attempts to build space nuclear reactors for these purposes. Despite this extended effort, neither NASA nor DOD has deployed an operational SNP system. It is now ap-

Table 1. Key Characteristics of Solar, RTG, and Fission Reactor Space Power Systems

	Solar	RTG	Fission Reactor
Energy Density (mass-volume-output)	Low	High	High
Electrical Power Level	Up to High	Up to Moderate	Up to Very High
Reliability	High	Very High	High
Propulsion	No	No	Yes
Life	Long	Very Long	Very Long
Cost	High (1)	Low	Low (2)
Relative Safety	Very High	Moderate	High (3)
Publicly Perceived Safety	Very High	Low	Low
Output Decreases with Distance from Sun	Yes	No	No

(1) A large part of the cost of a solar-powered space system is incurred in the launch budget. This is because solar panels are heavy for their power output. Further, their life is relatively short (about seven years, sometimes longer) and demands a frequent launch cycle to maintain a satellite constellation.
(2) The cost of a developed reactor is relatively low, and in assigning this rating we have not included the initial development costs to produce a space-rated reactor. This was done to make a baseline comparison against the other technologies presented, whose initial development costs are already met. Further, the cost for a fission reactor is only low if the developing nation already has a nuclear program to provide fuel.
(3) Assuming the reactor is activated outside the Earth's gravity well (about 1,000 km orbit).

proaching 50 years since the initial proposal of SNP. It is difficult to think of another scientific development program that political concerns have stalled for so long. Therefore, to contemplate a modern public policy maker's decision as to whether or not to deploy SNP, we must understand both the scientific history of SNP and the politics that have delayed the deployment of reactors in space for so long. This section highlights the key points that one can deduce from the history. Appendix A presents a more detailed review of SNP past programs.

Before moving to a brief history of SNP, the authors offer RTGs as an example of another nuclear-based technology used for several space missions. RTGs are made using an isotope of plutonium (Pu-238). This isotope generates large amounts of heat as it undergoes radioactive decay, and the heat is then converted to electricity. RTGs have been the mainstay for virtually all of NASA's deep space missions because of their high reliability and long operating life. Until the 1980s, many NASA missions used RTGs without incurring serious political opposition. However, in recent years opposing voices have increased in volume if not number. They are also more organized and politically effective than before. The most significant example of public protest against an RTG-SNP system was directed at the *Cassini* mission that NASA launched in 1997. Appendix A discusses the *Cassini* mission and its political impact.

While RTGs are not the primary subject of this paper, the political conditions of the recent past have made it clear that continued use of RTGs might be difficult. In fact, the political problems associated with RTG-based systems spill over to the debate about reactors, if for no other reason than that they both use nuclear materials as their primary source of energy.

Both the scientific and the political history of SNP shape the current public policy debate. The early period of SNP occurred during a time when scientists enjoyed tacit if not explicit permission from society and the government to develop SNP. In addition, nuclear power projects were supported and promoted by popular culture that included the entertainment industry and extended to advertisements in the popular press. The early programs were a scientific success and proved the technical feasibility of SNP. The question is, given the initial successes and the cultural permissions to pursue SNP technology, why has so little progress been made to date? To answer this question, one must understand the linked influences of science and politics within the historical context.

From the 1950s through the late 1990s, there was no compelling requirement for power in space that an alternative technology could not provide. Although SNP is technically feasible and extraordinarily capable in comparison to alternative technologies, the politics of nuclear technologies appears to have been a major impediment to SNP. This is apparent when SNP is compared to the naval nuclear submarine program. It is also clear that SNP quickly became a victim of the general fear and anxiety that ground-based nuclear power caused in the minds of the public starting in the early 1970s. The risks and rewards of SNP are in the public domain, and the political environment has made it extremely difficult to move the program off the drawing board and into space.

Complicating the situation is the political legacy of nuclear technologies from the Cold War. The Cold War gave governments, particularly those of the United States and the Soviet Union, essentially carte blanche authority to develop nuclear technologies for the national defense and then to extend their application into the civilian arena, notably for electrical power generation and medicine. The Soviet Union did it through communist central authority. The US government had tacit approval for SNP from its public because of overriding national security concerns. The US population had yielded to what sociologists call the "authoritarian reflex," that is the tendency of populations to trust government when they feel threatened.[3] Accordingly, the people trusted their government to wisely use a technology that few properly understood.

When the Cold War ended with the fall of the Berlin Wall in 1989, so in large part did the nuclear security imperative. With the security imperative gone, and with the rise of individualism and postmaterialist values in the latter third of the twentieth century, the US population effectively withdrew carte blanche permission for the government to do as it would with nuclear technologies. The public disclosure of some of the risks both the US and Soviet governments took in experimenting with nuclear technologies in the 1940s and 1950s gave many pause, as did the 1979 Three Mile Island and 1986 Chernobyl nuclear power station accidents. Emergent disclosure about those early experiments and the public risk they engendered exposed a new concern for policy makers: transscience.[4]

Transscientific Policy

Transscientific policy concerns those questions that arise from science that science cannot directly answer. Researchers are unable to experiment either because of limitations in technique or because experimental outcomes might involve excessive risk. There is a temporal dimension to transscience as well, just because an issue is transscientific today does not mean it necessarily will remain so as scientific technique improves or the political environment changes.

Nuclear technologies were the first instance of a scientific problem that entered the political realm as transscience. If mishandled as a matter of public policy, negative consequences of the nuclear experimentation could be momentous. For example, when Enrico Fermi was looking into the possible consequences of detonating a nuclear explosion for the Trinity experiments in the 1940s, one hypothetical outcome was that the nuclear explosion would start an atomic chain reaction in the Earth's atmosphere. The atmosphere would be burnt off, destroying life on Earth. Fermi himself went to the president to report this conclusion and his concern. The possibility of such a catastrophic outcome was later discounted through calculation and limited experimentation. However, there was never total surety, and the successful conclusion of the Trinity event must have brought tremendous relief. Transscientific doubt concerning risk coupled with the possibility of catastrophic consequence accompanied the first blaze of nuclear science. For some uses of nuclear technologies, such as SNP, transscientific doubt about risk and consequence still exists.

In today's political environment, SNP falls squarely into the category of transscientific policy because of some of the risks it inherently entails. For example, the risk of inadvertent release of Plutonium 238 from an RTG over a wide area due to a space-system mishap could have environmental consequences that cannot be safely tested, although it can be inferred. Such inferences are open to dispute among experts. Receiving conflicting expert advice leaves both policy makers and the general public in a quandary as to whether SNP programs should be supported. Consequential risk analyses for particular events can easily bridge many orders of magnitude depending on which expert speaks and how they weigh empirical factors.

In the transscientific context, decision making becomes a matter of judgment, with the most salient concerning the risk inherent in the enterprise—risk both measured and perceived. For a particular program, risk should be considered in concert with reward and relativity, as represented in figure 2. These three axes inside a transscientific context represent the foun-

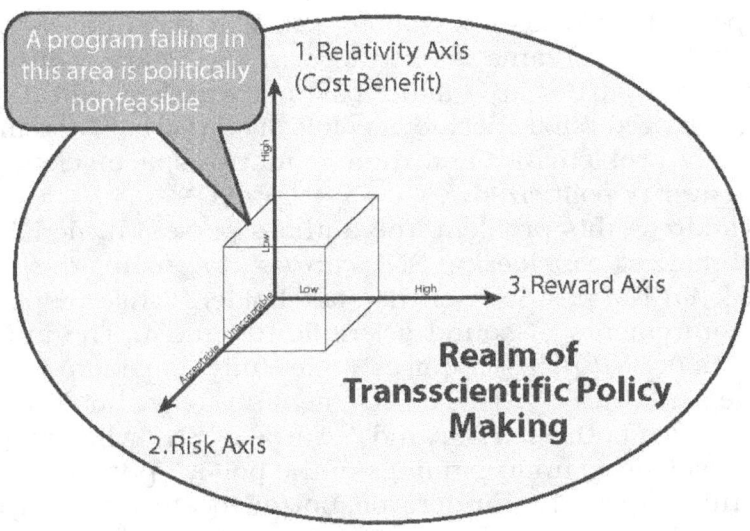

Figure 2. Space nuclear power program feasibility.

dation of the policy model the authors propose in chapter 4 as an aid to decision makers needing to make political judgments about transscientific programs, such as SNP programs.

The intent of this paper is to address the question: What mechanism would improve the political feasibility of a nuclear power program for US space operations? The authors' inquiries have highlighted the fact that the answer to the question is highly contextual and mainly a matter of political judgment. Unlike pure science, transscientific policy must include scientific data interpretation by inference and political value judgments. Transscience is the art of synthesizing political solutions that science informs but cannot solve. Empirical analysis is a necessary but insufficient tool for solving transscientific problems. That conclusion, and the fact that some stakeholders have not assimilated it, causes serious problems in engaging the public with respect to SNP. The result is the discomfort felt even in politically moderate circles. One side talks about empirical scientific facts (the proposing agencies) and historically has largely ignored the public face of the political debate. The other side counters with environmental and socially derived values (the public opposition), focusing on worst-case scenarios and potentially disastrous outcomes. The potential value of SNP-enabled programs is sacrificed in the name of ultimate safety. Both parties are talking past one another, and the ensuing polarized public debate is politically divisive. SNP remains politically problematic, and the conduct of space science remains overtly politicized.

To address this problem, the authors present a model that policy makers considering SNP can use to ensure they have considered the positions of the stakeholders while respecting the requirements of sound scientific judgment. The authors argue that SNP policy requires a carefully considered public engagement strategy. This engagement strategy should inform the public of both the risks and rewards of SNP while respecting the scientific underpinnings of the policy options.

In this paper, the authors do not advocate for or against SNP. In the context of space missions, SNP is an enabling technology that needs consideration in the risk, relativity, and reward framework proposed. SNP has its own advantages and limitations that, when considered in conjunction with the political context, will either enable its use or not. For the SNP decision maker—who in this context is finally the president, as the approving authority for nuclear space missions, supported by Congress, as the body that authorizes funding—SNP lies at the intersection of technical risk, the public's perception of that risk (social risk), and political risk. We advocate a policy model that, if wisely exercised, will improve the NASA or DOD program proponents' chance of having a rationally based program seen as politically feasible by the decision makers. We argue that such feasibility is contingent on the weight of public opinion being, at worst, insufficient to stop the program. The purpose of the policy model is to outline a democratically legitimate and scientifically sound mechanism to assist policy makers in considering SNP as a transscientific policy problem.

As illustrated in figure 2, the aim of the program proponent, in exercising the authors' policy model, must be to minimize the area of program "nonfeasibility." Further, the program proponent should have developed measures of effectiveness (MOE) to satisfy decision makers of the probity and political feasibility of his proposal.

In approaching the research question, the authors discovered that little has been written on the subject of SNP from a policy perspective. This is somewhat surprising given the long history of SNP. Numerous volumes contain the technical plans and details for SNP systems and their applications. However, the lack of policy analysis indicates a need that this paper seeks to satisfy. In that vein, the arrangement of the remainder of the paper to present the arguments outlined is as follows:

- Chapter 2: "Space Nuclear Power as Transscientific Public Policy," describes the nature of transscientific enquiry and relates it to SNP programs.

- Chapter 3: "Political Permission—the Contemporary Dimensions," examines contemporary society and its relationship to technology and science to draw out the implications SNP raises for contemporary policy and decision makers.

- Chapter 4: "A Transscientific Political Engagement Strategy," presents the values-focused decision strategy in association with the authors' policy model.

- Chapter 5: "Conclusions," summarizes the authors' research approach and the authors' recommendation for the use of a values-focused decision strategy to determine the political feasibility of transscientific policy options.

Notes

1. NASA, Jet Propulsion Laboratory, http://www.jpl.nasa.gov/jimo/ (accessed 5 Mar 2004).

2. NASA, http://spacelink.nasa.gov/Instructional.Materials/NASA. Educational.Products/International.Space.Station.Solar.Arrays/ISS.Solar. Arrays.pdf (accessed 5 Mar 2004).

3. Joseph S. Nye, Philip D. Zelikow, and David C. King, eds., *Why People Don't Trust Government* (Cambridge, Mass.: Harvard University Press, 1997), 218.

4. First described by Alvin M. Weinberg, *Science and Trans-science* (Alexandria, Va.: Minerva, 1972).

Chapter 2

Space Nuclear Power as Transscientific Public Policy

Many of the issues which arise in the course of the interaction between science or technology and society . . . hang on the answers to questions which can be asked of science and yet which cannot be answered by science.

—Alvin M. Weinberg

This chapter examines SNP as transscientific public policy. From the scientific and historical perspectives, SNP is hypothetically possible and technologically feasible. Yet, from the standpoint of weighing the political risks and rewards of pursuing SNP, science alone cannot and should not determine public policy. Political concerns about unavoidable public risks, potential consequences, and political rewards proscribe scientists from proceeding with experimentation and observation that would objectively verify the risks and rewards of SNP as public policy. These political limitations on the conduct of nuclear science place SNP in the realm of transscientific policy making.

SNP, as is often the case with transscience issues, generates political controversy by the intermingling of untested scientific hypotheses with competing social values about the political risks and rewards of a policy option. In the case of SNP, political controversy primarily comes from the perceived risk in launching, orbiting, or deorbiting fissile material. As with most transscientific issues, the science that initially created the political option continues to inform the political debate. However, scientific methods are inherently apolitical and not well suited to make the value discriminations required in political judgments. In a democratic society, scientific methods alone cannot resolve transscientific political controversies.

The Political Limitations of the Nuclear Sciences

This section examines the nature of normal apolitical science to determine how the continuation of the nuclear sciences can become politically infeasible. Thomas Kuhn, in his seminal work *The Structure of Scientific Revolutions*, describes science as a process that develops iteratively and is one in which anomalies in the scientific community's world view accrue until that world view must be fundamentally altered. Kuhn describes normal science as follows: "Normal science means research firmly based upon one or more past scientific achievements, achievements that some particular scientific community acknowledges for a time as supplying the foundation for its further practice."[1]

Therefore, normal science can be considered a strategy for choosing what to believe about the natural world. In the case of SNP, research and achievement in space exploration and the nuclear sciences are the basis for political policy options. These scientific achievements have provided a framework of beliefs and knowledge that permit the formulation of scientific hypotheses that in turn create new public policy options, such as placing nuclear reactors in space. However, SNP has political repercussions that scientific methods alone cannot address. SNP is a public policy option that cannot be scientifically tested or verified a priori because the potential political risks and consequences make scientific rigor impracticable or unethical to attain without proper political oversight. Conversely, respecting the rigorous requirements of sound scientific judgment is necessary to achieve proper political oversight.[2]

In transscientific policy making, science and politics are inextricably linked. When considering SNP as public policy, researchers must share the risks and rewards publicly and cannot confine them to a laboratory environment. Researchers must respect the governance process before space nuclear experimentation can take place. Science provides the policy maker with powerful tools that permit selection, evaluation, and criticism of SNP policy options. However, science does not provide the tools necessary to make political judgments concerning competing social values or to evaluate what risks are politically acceptable in a democratic society.[3] Even within the scientific community itself, social values often vary considerably from one discipline to the next.

These values inevitably affect scientific data selection, experimental design, outcome interpretation, and the criticism of scientific technique. Social values and public perceptions of risk and reward already have a powerful effect on ongoing scientific research and public policies, and especially the public policy options that nuclear science creates and informs.[4] Ultimately, it is a political and not a scientific decision to accept or reject the risks of SNP, and science is but one of several sets of tools available to a policy maker for a political decision.

We now see that SNP has entered the realm of transscientific policy making because science posits that SNP is technologically feasible and highly desirable for many missions and at the same time informs us of the potential risks of moving fissile material into and out of space. Unfortunately, these political risks and rewards cannot be proven with sufficient scientific rigor before committing one way or the other on the policy option of placing nuclear power in space. Therefore, the political risks and rewards must be inferred from available information. The policy maker then faces the dilemma of determining whether the possible rewards of SNP outweigh the potential risks of launching fissile material.

The political debate over SNP is typical of transscientific public policy issues insofar as the limitation that causes science to move into the political realm is not an inherent limitation of the scientific method. Rather, it is an appropriate political limitation on the conduct of scientific inquiry in a democratic society. When the public risks of SNP cannot be confined to a controlled environment and the outcome of the endeavor has political consequences, then transscientific policy formulation is required. Ultimately, the goal of transscientific policy is to manage the political calculus of public risk and reward while protecting the intellectual integrity of the underlying science that enables the political option.

Political Actors and Transscience

Political actors such as the Florida Coalition for Peace and Justice (FCPJ) form communities that exert influence in a democratic society through their shared values. These values form the foundation of the group's political activity. In contrast with the skepticism of purely scientific researchers, the nature of political value judgments that scientific observation informs but does not verify frequently causes political actors to have rigorous and rigid perceptions of political good. This rigidity of belief is remarkable, since the underpinning scientific knowledge is by definition indeterminate or unverifiable in transscience.[5] Group dynamics also tend to reinforce political beliefs to ensure that shared beliefs exert a deep and enduring hold on the political actors' values.

Political actors often suppress fundamental scientific and empirical data because they are subversive of their basic political position. It is nevertheless important to note that a central premise of democratic politics is that the political community can discover what policies are best through open political discourse and rational public debate. Political actors concerned with transscientific issues will take great pains to defend the premise of open, rational discourse even while actively attempting to suppress scientific analysis or ongoing research. Therefore, transscientific political debate in a democracy tends to devolve into a strenuous and sustained attempt by the political actors to force public policies into conformity with the political value judgments presupplied by the actors.

Politics injected into the scientific process can result in politically biased interpretations of scientific fact and observation, and it is difficult and time consuming to reconcile the political actors' biases and differing observations. Transscientific issues tend to become seen increasingly as political issues and less scientific in nature as the prevailing group of actors consolidates its political victory. The prevailing group may also construct political barriers to scientific investigation to strengthen its political position.[6] The politicization of a transscientific issue can result in the long-term loss of opportunity for scientific investigation and research, and possibly, in the creation of an antiintellectual political atmosphere that stifles scientific inquiry.

Transscience and the Politics of Risk and Reward

This section defines the risks and rewards of SNP that nuclear science cannot verify due to political constraints. Consider reward as political advantage, or at least acceptance, to proceed with a policy option. Conversely, risk is the political liability of proceeding with a policy option. Together with the usual political tension of risk versus reward, consider transscientific policy making as involving the art of reconciling the requirements of science with the realities of democratic politics.

In other words, transscientific policy making has two distinct requirements. The first is to preserve the intellectual integrity of the scientific or technological inquiry that informs the public debate. The second requirement is to conduct policy making in an open and democratically legitimate way. Ultimately, transscientific policy making should develop sound public policies that eventually depoliticize science, even though the program itself is always subject to legitimate political constraints.[7]

Transscientific policy making can only offer methods for politically evaluating the technical aspects of SNP. Scientific inquiry informs politics of the possibilities of success or failure by casting light on the technical risks and consequences. Thus, transscientific policy formulation must articulate the political assessment of the possible technical successes or failures. Transscientific policy making must grapple with the political risks and rewards in a democratically legitimate way while preserving the intellectual integrity of the scientific endeavors underlying the political controversy.

Technical assessment of the transscientific risks and rewards of SNP are grounded in the hard sciences, such as mathematics, physics, and chemistry. However, in transscientific issues, the experimental tools of the hard sciences are limited in scope and precision by political concerns and resource limitations. Nevertheless, researchers can verify some aspects of SNP with scientific rigor. A great deal of data can be collected and observed without actually launching fissile material into space and much analysis can be done at the component level of SNP without public risk. Yet the overall soundness of

SNP as a public policy cannot be tested without positively committing to launching nuclear material.

The technical assessment of risk and reward consists of the following:

- technical feasibility assessment of SNP at the component level,
- evaluation of the technical risks of the policy option, and
- scientific extrapolation of the consequences of the policy option.

The political assessment of SNP risk and reward is grounded in the democratic governance process, and political assessments are highly sensitive to how the arguments are framed for public debate. Ultimately, policy decisions in trans-science depend on value judgments about political risks and rewards, which, in turn, arise from policy options that scientific methods cannot verify. Therefore, the efficacy of technical risk and reward analysis and the political acceptability of consequences are rarely agreed upon by the stakeholders.

The political assessment of risk and reward hinges on three issues:

- political will of the stakeholders,
- subjective evaluation of the political risks, and
- political consequences and opportunity costs of the possible outcomes.

This creates a classic political conundrum for the policy maker with respect to SNP. Nuclear technologies are profound technological enablers of space missions because they provide high levels of power over long periods with high reliability and comparatively low weight penalties. Nevertheless, not all the technical engineering difficulties can be worked out until the reactor is actually flown in space. The engineering difficulties draw attention to the fact that there are inherent risks in launching fissile material into space.[8]

These hypothetical risks, especially the environmental risks, cannot be assessed with technical certainty, and the possible consequences of a launch mishap make nuclear power in space politically infeasible without some sort of engagement strategy to minimize both the technical and political risks. An engagement strategy is needed that allows science to move forward and continue to inform the political debate and that eventually reconciles the available scientific data with divergent perceptions of political risk and reward.

This chapter established that SNP is a classic problem in transscience. In the modern context, policy decisions about SNP must consider this crucial aspect. The situation is further complicated by the contemporary context. Public attitudes toward nuclear power, particularly political permission for the government to pursue the technology, are not assured. The various aspects of the contemporary political climate are the subjects of the next chapter.

Notes

1. Thomas S. Kuhn, *The Structure of Scientific Revolutions* (Chicago: University of Chicago Press, 1962), 10–11.

2. For a more detailed discussion see Sheila S. Jasanoff, "Citizens at Risk: Cultures of Modernity in Europe and the U.S.," *Science as Culture* 11, no. 3 (2002): 363–80.

3. Sheldon Krimsky and Dominic Golding, *Social Theories of Risk* (Westport, Conn.: Praeger, 1992), 238.

4. Carlo C. Jaeger et al., *Risk, Uncertainty and Rational Action* (London: Earthscan, 2001), 82.

5. Kuhn, *The Structure of Scientific Revolutions*, 79.

6. For example see Karl Grossman, "Alternative Energy Meets Main Street," *New Age*, July/Aug 1999, 59.

7. For example, budgetary constraints could still arise without repoliticizing the science itself.

8. Patricia M. Stearns and Leslie I. Tennen, "Regulation of Space Activities and Trans-Science: Public Perceptions and Policy Considerations," *Space Policy* 11, no. 3 (Aug 1995): 182.

Chapter 3

Political Permission— the Contemporary Dimensions

Sociologically, there is a big difference between those who take risks and those who are victimized by the risks others take.

—Ulrich Beck
The Risk Society, 1992

The political context of SNP has changed fundamentally from that of the Cold War era. The Cold War focus of the United States and its allies on a monolithic, nuclear-capable external threat is diffused today. Consequently, the tacit and carte blanche public permission that the government had during the Cold War to pursue nuclear programs through executive decision is revoked today. That is especially true for programs that the public perceives to potentially pose significant environmental problems, such as SNP programs. As well, people today question authority far more than they did in the Cold War era; they trust government less than their forebears.[1]

All of this directly affects the political feasibility of SNP and poses a challenge to those proposing space programs where SNP is either preferred or necessary. As alluded to in the opening quotation, SNP program proponents no longer operate in a context where they can conduct the relatively simple technical assessments of risk—of which they are most familiar and use the conclusions to satisfy the government that their program is necessary and that risk is acceptably low. The new demand is that they justify their programs in the public arena. This means that programs must be politically and socially justifiable and supported, in addition to satisfying the normal internal programmatics.

The audience to which SNP proponents must present such justification and garner approval has broadened to the point where public interest groups of various stripes are influential in the decision process but are not formally part of it. In the case of NASA, it is plausible that the public is only just recognizing implications of the contemporary state of affairs with respect to SNP space science programs. NASA's problematic public engagement strategy relating to *Cassini* highlights this. In the case of DOD, the need for a national-security imperative and the perceived risk of linkage to space weaponization make SNP an issue of great public concern and a conduit for political activism.

This chapter examines the contemporary political dimensions of the problem SNP program proponents face in satisfying leery United States and international communities of the safety and practicality of SNP. More specifically, the chapter develops the contemporary dimensions of the SNP debate in terms of four areas:

- **the context**, with a focus on the contested nature of the environment, public perception of risk, and the rising hurdles of permission granted by the public to the government to employ SNP;

- **the imperative**, from the standpoint of program proponents: who proposes, why they propose, what they want, and how they seek it, with a focus on contemporary and near future missions where SNP is necessary or desirable;

- **the resistance,** by program opponents: who opposes, why they oppose, what they want, and how they seek it; and

- **the implications**, a summary of imperatives and implications leading to the policy analysis model presented in chapter 4.

The Context

Today's prosperous, stable, and democratic industrial society enjoys physical wealth and comfort arising from an economy founded at least in part on technology. However, societal polarization can result when people are simultaneously attracted to and repelled by the consequences of technology. This is particularly true for those technologies that the public perceives to intrinsically encapsulate both enormous social reward and momentous risk. Such technologies include the nuclear and genetic technologies, biotechnologies, and nanotechnologies.[2] To illustrate this point, figure 3 depicts the relative coolness the public feels towards those technologies perceived as high risk (genetic engineering and nuclear technology), compared to those perceived as benign (solar energy, computers, etc.). The assimilation of technical risk into a public gestalt has produced what Ulrich Beck described as "risk society" in his seminal work of the same name.

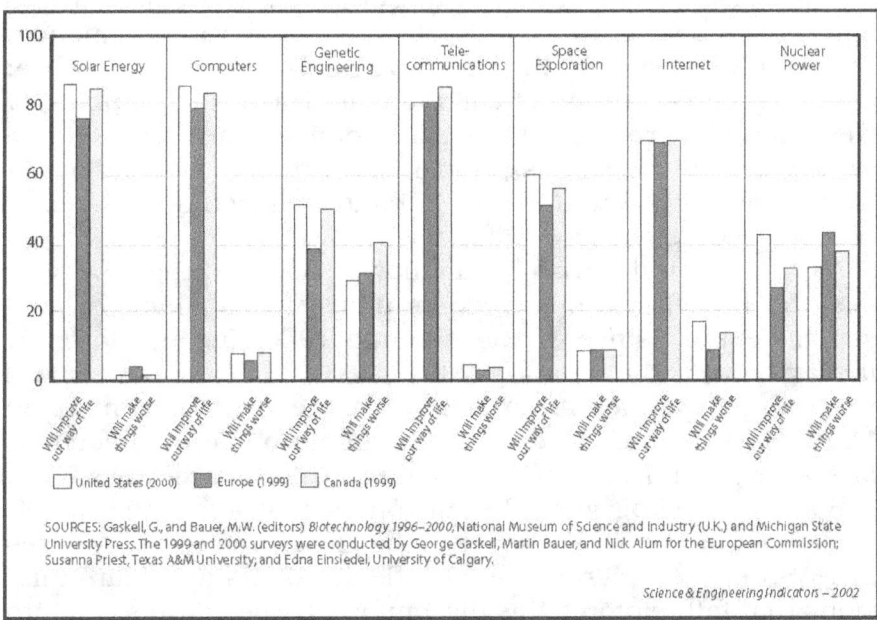

Figure 3. Public attitudes toward selected technologies, 2002. (Reprinted from National Science Board, *Science and Engineering Indictors, 2002*, figure 7-11, http://www.nsf.gov/sbe/srs/seind02/c7/fig07-11.htm [accessed 11 Mar 2004].)

A risk society is one that perceives risk in technology because it does not automatically trust it or its proponents, political or scientific, yet paradoxically enjoys technology's fruits, even if only those from a list sanctioned by their conscience. Some groups in risk society perceive cataclysmic outcomes from particular technologies or technology mixes and are prepared to protest against them. That is the case with SNP, where opponents claim, with some justification from the experience of the recent past, that the combination of a failure of a launch or orbital system carrying radioactive materials produces the equivalent of a dirty nuclear bomb. The reality of the risk is somewhat different, but such a perception exists in the public arena. So the battle is joined, SNP proponents versus opponents. But are they fighting in the same conflict space?

The conflict space is murky; it is essentially political, value laden, inhabited by the passionate on both sides, and covers an issue that is by nature transscientific. The object of the fight is the hearts and minds of nonaligned Middle America. The conflict space is further clouded by public ignorance: about 70 percent of Americans lack a clear understanding of the scientific process which compromises their capacity to understand technical risk and make rational, well-informed choices as to who to listen to in contested risk assessments, and why.[3]

As well, the public has been ambivalent about the costs versus the benefits of space exploration, with opinion divided roughly equally since at least the mid-1980s (fig. 4). However, it may be that the January 2004 presidential announcement regarding a manned lunar base as a precursor to a manned Mars mission—with an estimated cost of up to one trillion dollars—has galvanized public opinion regarding space exploration.

Society is feeling somewhat uncomfortable about the combination of nuclear and space technologies. Such recent failures in NASA space programs as shuttle accidents and Mars missions that fail reinforce this discomfort. These failures call into serious question the credibility of the scientific and policy-making elites and their ability to manage their programs safely. In addition, those championing postmaterialist values have seized upon the discomfort.

The Rise of Postmaterialist Values

Since the end of the Cold War in 1989, publics in prosperous, stable, and democratic industrial societies have become more physically secure, the upset of 11 September 2001 notwithstanding.[4] Their authoritarian reflex, or willingness to

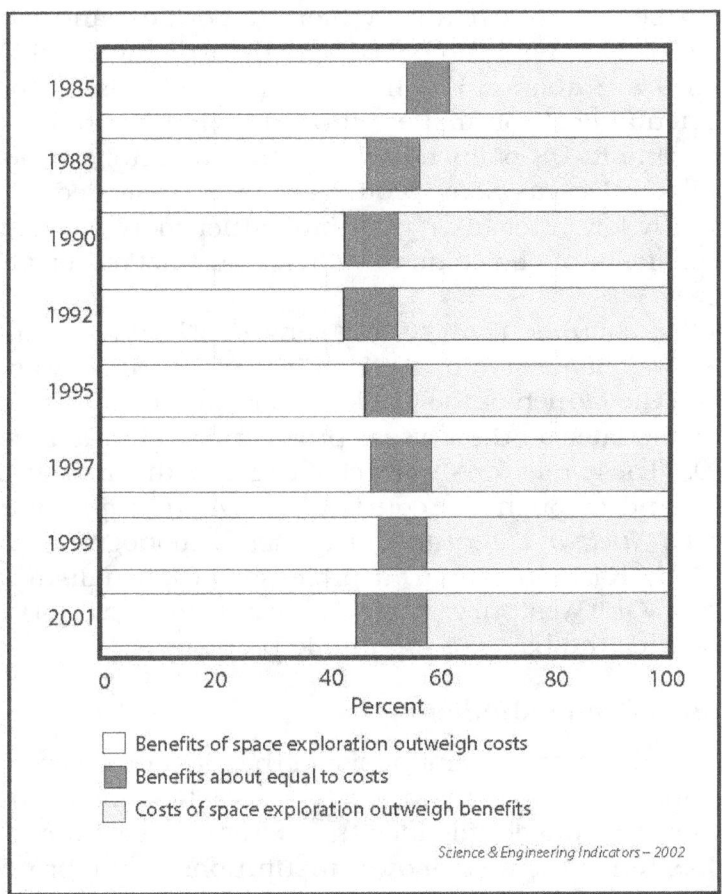

Figure 4. Public assessment of space exploration, 1985 to 2001. (Reprinted from National Science Board, *Science and Engineering Indictors, 2002,* figure 7-12, http://www.nsf.gov/sbe/srs/seind02/c7/fig07-12.htm [accessed 11 Mar 2004].)

acquiesce to government control, has diminished. Such societies exhibit high levels of dissatisfaction with their political systems and have less confidence in their political leaders and public institutions. Intergenerational stability in the provision of basic survival needs and comforts produces a postmaterialist mind-set where prosperity is taken for granted and society begins to focus on such other aspects of life as the quality of their and others political, social, and natural environments. Concern about these becomes relatively more important, and because postmaterialists tend to be educated, articulate, and well-read, they are able and willing to find mechanisms to steer governmental decision makers toward their preferred ends.

The reality in US politics today is that the public is not less politically active—as some would argue is evidenced through persistently low voter turnouts—but rather more so. However, the mechanism of choice now is often direct action, or political activism.

Significant public protests challenged SNP directly against NASA's *Cassini* mission of 1997.[5] Before that, NASA was challenged on the launch of the Galileo probe on the shuttle *Atlantis* in 1989 and later on the Ulysses probe on the shuttle *Discovery* in 1990.[6] These missions were challenged both through public activism and through the courts. *The Wrong Stuff: The Space Program's Nuclear Threat to Our Planet*, a monograph written in 1997 by Karl Grossman, a professor of journalism at the College of Old Westbury, State University of New York, best encapsulates the basis of the public protest.[7]

The Rise of Individualism

The balance between emphasis on the individual versus emphasis on the community in Western societies has been moving toward the individual since the industrial revolution.[8] This trend has tended to disempower institutions in the broad, not just political intuitions, and reached a nadir in the United States in the 1960s with the youth revolutions. In 2003 many of those same youth are now adult postmaterialists. In the 1960s the United States underwent a "Great Awakening" that created a central paradox. Society directly challenged institutional authority while at the same time demanding that government deliver the social services desired and reflect a social conscience. People seemed to want government to be everything and nothing: service provider and social shaper, but without intrusion into their lives or a high dollar cost. These conflicting demands have produced "a regime of activist government and activist anti-government politics that they [the people] can little understand, much less sense they are controlling."[9]

Underpinning the trend to individualism and activism, and the ensuing public confusion they have created, is the most widely accepted driver of human behavior in social science, self-interest.[10] Self-interest can be expressed in many dimensions, including those that appear altruistic such as environmentalism, but altruism in outcome should not be mistaken for altruism in motivation.

The realpolitik effect of the rise of postmaterialist values and individualism in the United States is that governing is harder. Less authority is ceded to the government, as is less respect to public institutions generally. Government policy often has to be transacted directly with issues-based groups, which may or may not represent the broader interest or hold the most balanced view, but must be accommodated to some degree. The degree of accommodation awarded to any particular group by the government is a political judgment, and perhaps more often a factor of media than merit.

Contested Ground

The nature of the political conflict around SNP is essentially polarized, as conceptually depicted in figure 5. On one side are the program proponents, which in the case of SNP are represented by government agencies. These include NASA, whose interest is space science, and DOD, whose interest is terrestrial security using space-based systems in a support role. These principal proponents are supported by a small constellation of special interest groups, both national and international, seeking to promote SNP for their own ends or interests. For example, various man-to-Mars factions support NASA.[11] For the opposition, the largest anti-SNP protest, that against *Cassini*, was led by the FCPJ, with its own constellation of allied actors.[12] With SNP, as with most issues brought into the public arena by proponents and opponents, the bulk of the public is neutral unless activated.

What both of the SNP polar groups want is political victory such that their preferred outcome is achieved. If they are wise and not totally desperate for their way, each will also seek a victory via methods that do not destroy trust. This means that in pursuing their ends, both sides would choose means that

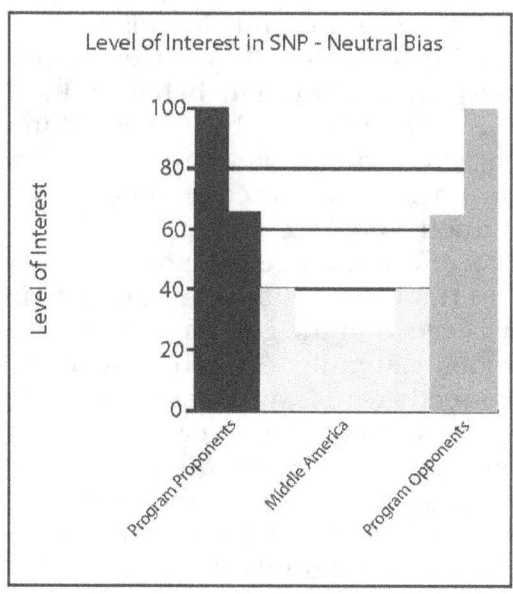

Figure 5. Societal level of interest in SNP: Illustrating neutral bias.

are deemed socially effective and morally sustainable over the long term. Otherwise, it is a case of "win the battle and lose the war."

To obtain their political victory, each side will seek to raise the profile of their argument to mobilize Middle America to their side (as conceptually depicted in figures 6 and 7), or at the least have the other side fail to mobilize Middle America against it. There is also a virtual aspect to any such engagement. Because each side will wage the public part of the battle largely through the media, it may be that generating the illusion of public alignment, by generating media favorable to one side's case, is enough to tilt the political balance.

Historically, Middle America has been the contested ground that program proponents and opponents have fought over. The null or negative position for either group is to use an effective public engagement strategy to keep the other side from activating and aligning at least part of Middle America to their end. If you are the program proponent, you also will want to keep your program in a favorable light with the political decision

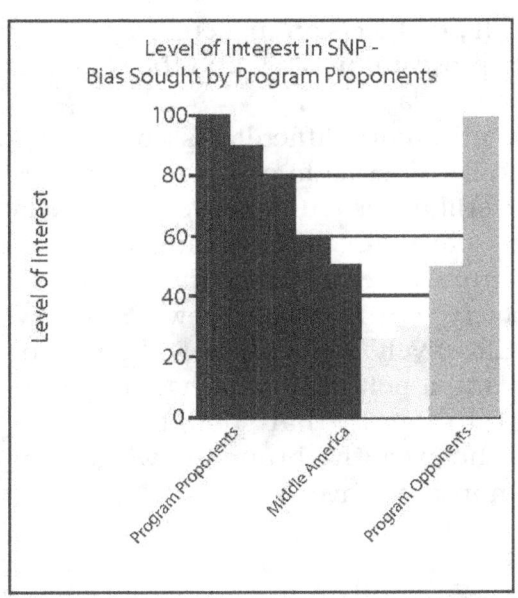

Figure 6. Societal bias and activation sought by program proponents.

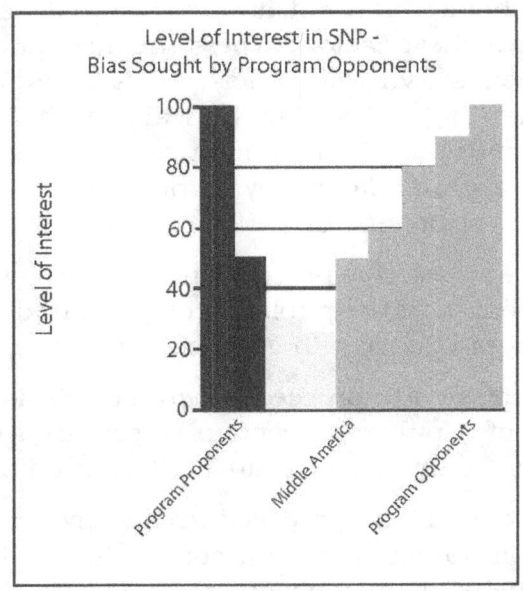

Figure 7. Societal bias and activation sought by program opponents.

makers. For SNP, the decision makers are Congress, the executive branch for program funding, and the president for a launch of fissionable materials.

The positive and more difficult task of activating and aligning part of Middle America to one side's ends requires domestic and, better still, international leverage. The logic from both sides is that a victory—in either halting an opponent's strategy or strongly aligning at least part of the normally neutral stakeholders in society to a favorable view—is a powerful political mechanism. Effectively, the victors will gain public avowal of their position and a political following, or at least the absence of a disavowal. In theory, that should leave the way clear for Congress and the executive branch to weigh the political risks of the program in their favor.

Political Will

The aim of SNP program proponents and opponents alike is to convince the political decision maker to their end, or at least to an acceptable compromise. But the government is not a neutral stakeholder, and not always a supplicant subject to the whim of whichever actor garners the most public support. The government's own will, beliefs, grass roots support, and strategies regarding an issue are critical factors that shape the strategies of both proponents and opponents if they are wise.

In pursing a particular policy or program in a democratic system, the government can work from one of three premises:

- explicit permission to proceed enabled through successful engagement with the public through shared process that generates public trust in government;

- tacit permission to proceed, where permission is granted because of a public perception of serious threat or vulnerability, so invoking the authoritarian reflex;[13] or

- autocratic decision to proceed irrespective of any consensual public process, with a concomitant willingness to accept the political consequences, as with Pres. John F. Kennedy pronouncing that America would go to the Moon

(1961), or Pres. George W. Bush pronouncing that America would go to war with Iraq (2003).[14]

From the list presented, the mechanism chosen by government to affect a program is a question of political will and judgment. The public perception of risk, relativity, and reward shapes but not necessarily controls this choice.

It is fair to say that, given the nature of American postmaterialist society, unless there is an overwhelming and publicly understood imperative for a particular program, which is not currently the case with SNP, obtaining explicit permission from the public is generally the most effective course for program proponents to pursue. This is because the public sees process and policy fairness as important.[15] That is particularly true for programs that have long time frames covering several election cycles, and where widely understood and supported outcomes of social value would not exist without explicit, effective public engagement. Programs incorporating SNP are of this ilk.

International Treaty Restraints

Two international treaties and one export control protocol have governed space policy for the last several decades. They are the Outer Space Treaty (OST) of 1967, the Antiballistic Missile (ABM) Treaty of 1972, and the Missile Technology Control Regime (MTCR) of 1987. Briefly, the OST states that international law applies beyond the atmosphere. The treaty reemphasizes that, in accordance with the 1947 United Nations Charter, one sovereign state cannot threaten the territorial integrity or political independence of another. The OST initiated new space-related agreements aimed at assuring free access to space and celestial bodies. Additionally, it prohibits national appropriations of space or celestial bodies, and finally, prohibits placing any weapons of mass destruction (WMD) in space or on celestial bodies.[16]

The ABM treaty prohibited the development, testing, or deployment of space-based ABM systems or components. It limited the United States and the Soviet Union to a single terrestrial ABM site each with a maximum of one hundred missiles. It also prohibited interference with the national technical means (that are, in part, reconnaissance satellites) used to verify treaty compliance.[17] On 13 June 2002, the United States withdrew from the ABM treaty citing national security concerns. This freed the United States from the treaty prohibition against testing or deploying weapons in space other than WMD.

The MTCR is an export control regime signed by the leading space-faring nations. Its purpose is to prevent proliferation of rocket technologies beyond a closed circle of countries already possessing them.

For the purposes of this paper, the important consideration is that international treaties of themselves are not obstacles to a responsibly managed SNP program. The extension of conflict into space and perhaps the weaponization of space are different matters entirely, but those are not the subjects of this paper.

The Authoritarian Reflex

Despite the difficulty that current societal attitudes cause those who would govern, the people will cede control to their government, and tend to trust government more, when they feel threatened. This behavior is the authoritarian reflex as illustrated in figure 8.

Figure 8 illustrates the high level of trust US citizens placed in their government in the 1950s and 1960s. This trust was a product of the social mores of the day and the level of threat felt by society. The trust placed in government afforded it carte blanche to develop nuclear technologies in the first two decades of the Cold War. The figure also illustrates the emergent decline in trust in the early 1970s, which arose primarily as a result of the Vietnam War and also because of the postmaterialist factors already discussed. What is interesting is the spike in trust that followed al Qaeda's attack on the United States on 9/11, and its brevity. However, the spike demonstrates that even today, people will cede more authority to the government when they feel threatened. Thus, it highlights the possibility that a well-engineered and relatively safe SNP-based program could gain public acceptance should the security environment deteriorate in the future.

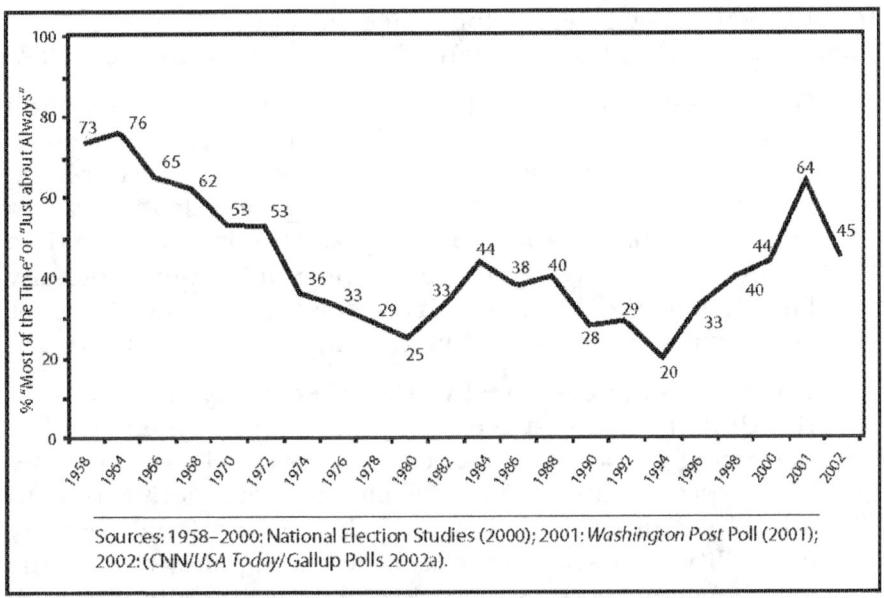

Figure 8. Citizens who trust the United States federal government, 1958 to 2002. (Reprinted from David King and Zachary Karabell, *The Generation of Trust: How the United States Military Has Regained the Public's Confidence Since Vietnam* [Washington, D.C.: The American Enterprise Institute Press, 2003], 2.)

The Imperative

The reality is that technology programs that are not tied to specific—and agreed upon—mission goals become very vulnerable to budget cuts or even cancellation over time.

—Cong. Bart Gordon
Ranking Democrat
US Congressional Sub-committee
on Space and Aeronautics, 2003

Within the United States, there are only two government agencies capable of, and currently interested in, pursuing SNP for contemporary and near future missions. They are NASA and DOD in partnership with other national security agencies. The first chapter presents the underlying technical reasons as to why SNP is of interest for space missions. Three generic missions linked to their stakeholders are candidates for SNP.

- **Space and space-based science missions.** In the United States, space science is the domain of NASA, which is supported by scientific advocates of SNP. These advocates believe that nuclear power is the only practicable way to develop a sustainable human presence in space and to conduct space research where the Sun's energy density falls below the level at which solar power is viable or where solar panels would be too large or heavy for launch.

- **Space and space-based national security missions.** In the United States, space-based security missions are the domain of US security agencies, supported by national security proponents who advocate a more active role in space because they believe it enhances national security capability with acceptable technical and policy risk. Some also believe that acting quickly to seize the ultimate high ground of Earth orbit to support terrestrial operations will confer a "first mover" advantage outweighing the potential disadvantages to national security and international stability making space an arena for direct action, rather than the support environment it is currently.

- **Emergent missions.** Emergent missions will be a product of the first two mission types, making SNP de rigueur in government-run space operations. From a proven base, others may wish to move SNP technology laterally into new applications. Commercial missions would benefit from the technology, as would other nations to support their own interests. Over time, commercial missions could emerge parallel with the development of US national capabilities, but this is unlikely because the risk, cost, and national control of nuclear technologies would require a fundamental policy shift as to how nuclear materials are controlled and who operates nuclear systems. Alternatively, and more likely, a demand for commercial operations could arise at an opportune time in the future. The time frame for an emergent demand for commercial SNP would be at least 15–25 years and as such is not for consideration in this paper. However, the issue of SNP proliferation by existing nuclear states could potentially emerge, as could counterspace concerns, and these are germane.

Of the listed mission categories, the first two arise from US space policy.

US Space Policy

Pres. William J. Clinton signed the most recent US National Space Policy in 1996. National Security Presidential Directive/NSPD–15, signed by Pres. George W. Bush on 28 June 2002, calls for a review of this policy. Until such a review is complete, the 1996 National Space Policy outlined by President Clinton remains current. The Clinton era document essentially "repackaged the same goals summarized in the National Space Act of 1958 and updated them in the context of current commercial and military landscapes."[18] NSPD–15 stresses that "access to and use of space is central for preserving peace and protecting United States interests."[19] Though not citing specific threats, the policy states that the United States will conduct those space activities necessary to ensure national security, which include "assuring that hostile forces cannot prevent our own use of space" and "countering, if necessary, space systems and services used for hostile purposes."[20] It goes on to state the goals of the national space agenda, five of which are

- knowledge by exploration,
- maintenance of national security,
- enhancement of competitiveness and capabilities,
- private sector investment, and
- promotion of international competition.

DOD space policy focuses on operational capabilities that enable the military services to fulfill the national security objectives. It enumerates three space-related tasks that guide the military services: (1) deter or, if necessary, defend against enemy attack; (2) enhance the operations of US and allied forces by employing space systems; and (3) ensure that forces of hostile nations cannot prevent US use of space. In 1998, the secretary of defense's 1998 annual report asserted that space power had become vitally important to the nation for economic as well as military reasons.

> The world is increasingly transitioning to economies in which information is a major engine of prosperity. While United States national security interests focused in the past on assuring the availability of oil, the future may require greater interest in protecting and accessing the flow of information. As a result, the importance of space as a principal avenue for the unimpeded flow of information throughout a global market increases. DOD recognizes these strategic imperatives and will assure free access to and use of space to support United States national security and economic interests.[21]

The report goes on to state that the United States should anticipate attacks against US and friendly allied space systems in the future and declares that, "DOD must have capabilities to deny an adversary's use of space systems to support hostile military forces."[22]

The Air Force Space Command (AFSPC) has taken the above policy and guidance and crafted a strategic master plan. Its vision statement declares that AFSPC will develop "[a] globally integrated aerospace force [capable of] providing continuous deterrence and prompt engagement for America and its allies . . . through control and exploitation of space and information." AFSPC capabilities will

> enable a fully integrated Aerospace Force to rapidly engage military forces worldwide. Our space forces will move beyond being primarily force multipliers to also being direct force providers. Global real-time, situational awareness will be provided to our combat commanders through space based Navigation, Satellite Communications, Environmental Monitoring, Surveillance and Threat Warning, Command and Control, and Information Operations systems. Robust and responsive space-lift and improved satellite operations capabilities will provide on-demand space transportation on-demand space asset operations ensuring our ability to access and operate in space. Full spectrum dominance in the space medium will be achieved through total space situational awareness, protection of friendly space assets, prevention of unauthorized use of those assets, negation of adversarial use of space and a fully capable National Missile Defense.[23]

NASA and Space Science Missions

NASA has definite plans for space science missions that will necessitate SNP.

- **Project Prometheus.** In 2003, NASA renamed what was its Nuclear Systems Initiative as Project Prometheus. That reflected the sensitivity NASA has been feeling with regard to SNP and the new awareness within the agency regarding the importance of managing public risk perception through effective engagement, especially with regard to high-profile missions using nuclear power. Prometheus has three components:[24]

    — **The Jupiter Icy Moon Orbiter** has a planned launch date of 2012. JIMO is a flagship mission that will pioneer the use of a small nuclear-fission reactor for orbiter electrical power and indirectly generated thrust. SNP is the only current technology that will provide the power density necessary for the JIMO mission. To mitigate risk, current plans are that the reactor will not be activated until the orbiter has left the Earth's gravity well (about 1,000-kilometer low Earth orbit). Appendix B provides a full description of the JIMO mission.

    — **Nuclear Power** is an initiative designed to improve the efficiency of current RTG technologies.

    — **Nuclear Propulsion** includes the development of JIMO's nuclear reactor.

- **Manned Lunar Base and Man-on-Mars.** Pursuit of the manned lunar base and subsequent trip to Mars announced by President Bush in 2004 will necessitate the use of SNP in some form for technical reasons. Man will not be able to get to Mars without the advantages of power density and longevity offered by SNP over any foreseeable alternative power source. Of note, no government agency has mentioned the nuclear power aspect of the proposal, probably due to the public sensitivity of the issue.

DOD and Others and Space-Based Security Missions

At the time of writing, DOD has no known plans on the table to field a space-based system reliant on nuclear power. However, a 2001 Space Commission report, chaired by the current secretary of defense, Donald Rumsfeld, calls for the United States to maintain superior capability both in terrestrial operations supported from space and in space operations themselves.

> We know from history that every medium—air, land and sea—has seen conflict. Reality indicates that space will be no different. Given this virtual certainty, the United States must develop the means both to deter and to defend against hostile acts in and from space. This will require superior space capabilities. Thus far, the broad outline of United States national space policy is sound, but the United States has not yet taken the steps necessary to develop the needed capabilities and to maintain and ensure continuing superiority.[25]

The report also emphasizes the necessity for superior US space-based intelligence, surveillance, and reconnaissance, and space control. While missions envisioned under these drivers do not absolutely need SNP, any simple analysis demonstrates that they would benefit by using nuclear power because of its intrinsic advantages. Direct costs would fall, and mission effectiveness would be enhanced by a small, light, compact, long-lived system that provided both ample electrical power and thrust for on-orbit maneuver.

If DOD enters the SNP business, though, a major shift will occur. Instead of the few, rare SNP system launches that NASA would execute primarily for deep space missions, DOD missions would necessitate that SNP operations become commonplace. It is easy to envision constellations of nuclear-powered satellites in orbit. One candidate system would be advanced space-based radar, or perhaps a system with a mix of active and passive hyperspectral sensors. Regular SNP operations in low Earth orbit would add a new dimension to the public's perception of risk. They would also require a different imperative to establish an effective public engagement process about the political feasibility of SNP before there is a need to increase the number of operational SNP platforms. In addition, SNP systems in orbit around the Earth will certainly be of concern to numerous international stakeholders.

Overall, it is fair to say that the only reason that DOD and others have not in the recent past sought reactor-based SNP systems is because of (1) the problematic political dimension, (2) the lack of a truly compelling mission that could overcome the political problem, and (3) the cost of developing such a system. However, if JIMO proceeds as planned, NASA will develop a reactor-based SNP system, providing an opportunity for the national security agencies to perhaps "hitch a ride" while allowing NASA to deal with the political issues under the guise of space science. So the fight is NASA's, unless there is an emergent security issue calling out for SNP that DOD is poised to exploit. NASA needs to confront committed opposition and win the public's hearts and minds to support a space-based version of those nuclear technologies they are wary of and uncomfortable with even in terrestrial applications.

The Resistance

What is interesting about SNP opponents is that they are not protesting the space operations per se, but rather the risk they perceive to be inherent in a particular enabling technology, SNP. In the case of *Cassini*, the 80 or so allied groups comprising the FCPJ have weighed the risk inherent in nuclear power with the fragility of space operations and have decided that

- the public reward of SNP does not warrant the risk, the reward being knowledge through space science in the case of *Cassini* and its forebears;
- NASA's SNP program has a direct link to nuclear weapons (a claim always denied by NASA and one that falls under the heading of popular conspiracy theory rather than substantial fact); and
- the FCPJ was determined to do something about it.[26]

To forestall NASA, the FCPJ's methods included public engagement, political activism, and formal legal action against the *Cassini* launch. FCPJ also gained attention by focusing on NASA's risk assessment, in particular the Final Environmental Impact Statement (FEIS). The FEIS is a necessary outcome of the 1969 National Environmental Policy Act (NEPA). NEPA requires that any major program conducted or authorized by the federal government be subjected to an environmental impact assessment. FCPJ challenged NASA's risk assessment methodology and conclusions claiming that both were incorrect and incomplete.[27]

The FCPJ is a fairly typical activist group. It has a small overhead with a full time staff of two in 1997 during its Stop *Cassini!* campaign. It is comprised of individual and organizational members. Most members are middle to high income, with the majority Christian, white, and college educated. Despite being able to attract up to 1,500 people to its protest rallies, and even with the legal action it took against NASA, the FCPJ did not believe it could stop the *Cassini* launch. However, according to its director, Bruce Gagnon, it could still provide an important vehicle for expression for people who otherwise felt powerless. Overall, the FCPJ is representative of a postmaterialist, issue-based group of concerned and responsible citizens as described earlier in this chapter.[28]

Cassini and its predecessors were launched despite the FCPJ's best efforts, but NASA seems to be coming to the realization that it is facing a capable and determined opposition whose values-based message has appealed more to the public than has NASA's own logic-based message. Further, if the FCPJ could enlist the support of more mainstream and universally appealing activist groups as Greenpeace and the Sierra Club, it is possible that NASA will lose a future fight for the hearts and minds of Middle America, making it politically difficult or impossible for the government to justify its future space plans on a risk basis, let alone a cost basis. JIMO and Man-to-Mars may be lost unless they are authorized through autocratic presidential decree. Such a decree, if contrary to the wishes of a hostile public, could come at high personal price to the president, and may or may not be pursued depending on the character of the administration of the day.[29] From DOD's perspective on the sideline, public opinion galvanized against SNP could cause the cancellation of Prometheus, with the subsequent loss of access to NASA-developed space-rated fission-reactor technology. That would make the space control advocates' longer term ambitions politically infeasible due to (1) prohibitively high startup costs, (2) an erosion of the technical expertise necessary to develop SNP, or (3) the fact that SNP in any form would not wash with the public.

This analysis leads us to where the United States is today. DOD is on the sidelines, perhaps waiting for a crisis of opportunity to obtain tacit public and hence political permission to meet its long-term ambitions in space. NASA, at least in some quarters, is seeking to improve its engagement with the public such that they can mobilize public bias to their ends. Where to next?

The Implications

Democracies of different sizes tend toward the traits presented in table 2.

Table 2. Traits of Democracies

	In smaller systems:	In larger systems:
1. The members are more	homogeneous	diverse
2. Incentives to conform to a uniform code of behavior are	stronger	weaker
3. In relation to the numbers holding the majority view in a conflict, the number who openly dissent are	fewer	greater
4. The likelihood that conflicts among groups involve personal conflicts among the individuals in each group is	higher	lower
5. Conversely, conflicts among organizations are	less frequent	more frequent
6. *Processes for dealing with organized group conflicts are*	*less institutionalized*	*more institutionalized*
7. *Group conflicts are*	*infrequent but explosive*	*frequent but less explosive*
8. *Group conflicts are*	*more likely to polarize the whole community*	*less likely to polarize the whole community*

Source: Robert A. Dahl and Edward R. Tufte, *Size and Democracy* (Stanford, Calif.: Stanford University Press, 1973), 92.

The issues of specific concern for SNP in the table are italicized. The United States as a large democracy aligns to the traits of larger systems. Dissenters can usually find enough allies in a large democracy to reach a threshold where dissent is effective, and that dissent can be persistent and overt.[30] This is the case for SNP, which points to the need to develop a public engagement strategy that forestalls community polarization.

NASA's logic-based arguments regarding SNP have not captured the public's hearts and minds. Although NASA has not had a launch stopped because of SNP activism, the swell of negative public opinion and the guarantee that the next launch will attract another debilitating legal challenge has given NASA cause to reexamine its communication strategy.[31] NASA seems to be coming to the realization that if it wants to capture Middle America as a foil to anti-SNP activism, public communication will have to be done differently.

What NASA did in the past, and to some extent continues to do, is frame SNP as a binary decision. A binary frame means that the public debate is forced such that it is over before it has begun, and the political decision makers are left to wonder how much political capital they will have to expend in support of the program, and whether that is warranted. The choice has been simple: either SNP enables the mission or the mission cannot be done. That SNP is needed for some missions is true, but framing the argument in these binary terms gives the public no latitude in the debate and allows program opponents to galvanize the public to only one possible position: NO. This from Professor Karl Grossman, author of *The Wrong Stuff*, when asked what could justify the use of SNP: "On the use of nuclear power in space, this may sound extreme but I don't believe it ever should be used above our heads—not in Earth orbit, not on deep space missions."[32] Of course, as a leading activist, Grossman will never change his mind. Nonetheless, leaving room to frame the problem in such a way as to meet the value-based preferences of the target audience (Middle America) would be wise and will enhance the political feasibility of SNP.

Some early offerings regarding Prometheus have been of the binary ilk. For example, in 2003 Lockheed Martin's vice president stated, "it is not whether we will discover new nuclear space power systems—but whether we will explore space."[33] The issue is not the necessity of SNP itself, but the framing of the technology.

Going Forward

This chapter establishes the contested nature of the current political environment and highlights the fact that contemporary society has withdrawn its carte blanche permission for nuclear technologies. NASA will carry the public policy debate for SNP forward, centered on JIMO, which is a compelling mission that requires the technology. DOD also has a strong case for SNP arising from emergent security concerns. The authors have stated that DOD should be establishing a mechanism to leverage such an opportunity. The authors have presented the case that NASA has not done well in convincing a leery public of the risk, relativity, and reward of SNP—as evidenced by its debilitating public debates with FCPJ thanks to an inadequate and antiquated public engagement mechanism.

In the next chapter, "A Transscientific Political Engagement Strategy," the authors present an alternative to the polarized binary public engagement framework NASA has been using. The alternative framework is one where both sides agree to seek a proposition based on shared values.[34] This allows public concerns to be brought out early and included in policy analysis and mission design. On the surface it might seem that this values-focused approach is applicable only to NASA. However, short of responding to an immediate national security threat, the approach also has value for DOD as it considers SNP for its own future space operations.

Notes

1. Joseph S. Nye, Philip D. Zelikow, and David C. King, eds., *Why People Don't Trust Government* (Cambridge, Mass.: Harvard University Press, 1997), 219–21.

2. The National Science Foundation has a large body of recent work available on the Internet on this subject. See http://www.nsf.gov/sbe/srs/seind02/intro/intro.htm.

3. National Science Board, *Science and Engineering Indicators, 2002*, chapter 7, "Science and Technology: Public Attitudes and Public Understanding," http://www.nsf.gov/sbe/srs/seind02/c7/c7h.htm (accessed 27 Jan 2004).

4. Nye, *Why People Don't Trust Government*, 219–21, 271.

5. An excellent master's thesis on the political and sociological aspects surrounding *Cassini* is by Victoria Friedensen, "Protest Space: A Study of

Technological Choice, Perception of Risk, and Space Exploration" (master's thesis, Virginia Polytechnic Institute and State University, 11 Oct 1999).

6. Patricia M. Sterns and Leslie I. Tennen, "Regulation of Space Activities and Trans-science: Public Perceptions and Policy Considerations," *Space Policy* 11, no. 3 (Aug 1995): 185.

7. Karl Grossman, *The Wrong Stuff: The Space Program's Nuclear Threat to Our Planet* (Monroe, Maine: Common Courage Press, 1997).

8. Nye, *Why People Don't Trust Government*, 271–72.

9. Ibid., 272.

10. Ibid.

11. The only way to put a man on Mars is to use SNP. Other power systems are inadequate for the mission.

12. See Appendix C for the list of actors comprising the FCPJ during the Stop *Cassini!* Campaign.

13. For example, the 9/11, terrorist attack.

14. There is one other mechanism for introducing a program, at least a national security program, and that is to run it secretly. Such programs are termed *black*, and the national security guise justifies their conduct in that making the program public would fundamentally compromise it. This paper does not discuss the black mechanism further, as it is outside the scope of the paper.

15. Carlo C. Jaeger et al., *Risk, Uncertainty and Rational Action* (London: Earthscan, 2001), 186.

16. Article XI and Article IX prohibit the weaponization or annexation of celestial bodies. A copy of the treaty is at http://www.oosa.unvienna.org/SpaceLaw/outersptxt.htm (accessed 25 Mar 2004).

17. The Federation of American Scientists (FAS) maintains a copy of the treaty, source documents, chronology, and news regarding the ABM treaty as part of their Weapons of Mass Destruction monitoring program, http://www.fas.org/nuke/control/abmt/ (accessed 26 Mar 2004).

18. US White House, *Fact Sheet: National Space Policy*, 19 Sept 1996, http://www.ostp.gov/NSTC/html/fs/fs-5.html (accessed 26 Mar 2004).

19. Ibid.

20. Ibid.

21. US Department of Defense, *Report of the Secretary of Defense to the President and Congress, 1998*, http://www.defenselink.mil/execsec/adr98/chap8.html#top (accessed 26 Mar 2004).

22. Ibid.

23. US Air Force, Space Command, *Strategic Master Plan for FY 04 and Beyond*, http://www.cdi.org/news/space-security/afspc-strategic-master-plan-04-beyond.pdf (accessed 26 Mar 2004).

24. Jeremy Hsu, *Project Prometheus: A Paradigm Shift in Risk Communication*, sciencepolicy.colorado.edu/gccs/2003/student_work/deliverables/jeremy_hsu_project_prometheus.pdf (accessed 8 Mar 2004).

25. US Department of Defense, *Report of the Commission to Assess United States National Security Space Management and Organization*, 2001, http://www.defenselink.mil/pubs/spaceintro.pdf (accessed 26 Mar 2004).

26. See Appendix C for a complete list of FCPJ members.

27. Friedensen, "Protest Space," 49.

28. Ibid., 44.

29. Ibid., 7.

30. Robert A. Dahl and Edward R. Tufte, *Size and Democracy* (Stanford, Calif.: Stanford University Press, 1973), 91.

31. We describe the legal challenge as debilitating because of the implications of program opponents timing it to this effect. Deep space science missions have particular launch windows that in most cases only open every few years. Therefore, a legal challenge to an SNP system launch only has to delay the launch past the launch window to disable the program. Programmatics are such that few programs can survive a several year hiatus. The implication is that the program opponents of SNP do not have to work to cancel a program; a well-timed legal challenge near the launch window will disable it, perhaps fatally.

32. Professor Karl Grossman to Wing Cdr Anthony Forestier, E-mail letter, subject: Policy and Nuclear Power in Space, 9 Oct 2003.

33. US Senate, Committee on Commerce, Science and Transportation, *In-Space Propulsion Technologies, Testimony by Mr. James H. Crocker, Vice-President of Lockheed Martin*, 3 June 2003. Quoted in Hsu, *Project Prometheus*, 6.

34. Joseph L. Arvai, Tim L. McDaniels, and Robin S. Gregory, "Exploring a Structured Decision Approach as a Means of Fostering Participatory Space Policy Making at NASA," *Space Policy* 18, no. 3 (Aug 2002): 221–31.

Chapter 4

A Transscientific Political Engagement Strategy

Science alone cannot establish the ends to which it is put.
—Francis Fukuyama

Achieving national goals in space will be extraordinarily arduous, both technically and politically, over the next few decades. There are numerous technological challenges that have inherent political dimensions. Some examples include proactively fostering cooperation on the international space station, ensuring the safety of flight personnel, and implementing technical performance plans as required by law. These activities take a significant toll on an agency's time and resources. Additionally, and perhaps most significantly, some programs, such as SNP, pose transscientific public health and environmental risks that affect the unscientific public and create political problems that can make program planning and implementation as demanding and complex as the science that originally enabled the project.

Responsibly implementing an SNP program will be a critical test of both technological and political skill. Barring an unforeseen breakthrough in fuel cell or solar technology, SNP will almost certainly be desirable for many future missions, and in JIMO's case, it will be essential.

Because the power source is the enabling technology for the conduct of all types of missions in space, planners must exercise special care when proposing SNP.[1] A political fumble in implementing SNP could effectively scuttle NASA's long-term plans for deep space exploration or manned expeditions beyond Earth's orbit. This outcome could also affect any future plans DOD might have for SNP.

Clearly, DOD would have different intentions for SNP than NASA and may be able to make a case for its necessity due to foreseen or emergent national security imperatives. However, DOD's political challenges are similar to NASA's. The underlying rationale for political permission is the only significant difference. Because SNP is a transscientific issue, an agency's engagement strategy must explain the technical risk as forthrightly as possible, meet the public interest politically, and protect the intellectual integrity of the science from undue political influence.

The history of reactor-based SNP is one of starts and stops. Without a publicly understood and politically compelling mission requirement, it is difficult to imagine any SNP program getting very far in the modern political arena, unless the president chooses to accept the political risk of an autocratic decision. SNP has significant advantages and is a generally well-proven technology. However, the perceived risk of its use in space exacerbates challenges faced by nuclear technologies. The decision strategy must recognize this history, and an effective engagement strategy will serve to minimize the real and perceived risks.

By using a values-focused decision strategy, one can construct a reasonable transscientific decision strategy that simultaneously respects the requirement for democratic legitimacy and scientific rigor in the policy-making process. The work of Keeney on values-focused thinking combined with current best professional practices in science and space policy making contributed to this strategy.[2] A values-focused decision strategy answers the political actors' empirical concerns using the best available scientific techniques and predictive tools. The strategy then uses objective scientific facts in the context of political judgments of risks and rewards to move policy forward democratically.

To implement a values-focused decision strategy for SNP, this chapter first outlines the classic three-dimensional policy components of risk, relativity, and reward pertaining to SNP. Next, the chapter examines these dimensions of the problem using a five-step values-focused decision strategy. This values-focused decision strategy has the advantage of maintaining scientific rigor where possible and yet retains the ability to examine the potential effects of science through a political lens where rigor is not possible.

The transscientific aspects of SNP make it impossible to objectively verify all of the stakeholders' concerns about the risk, reward, and relative merits of a particular policy option. However, researchers can craft a values-focused strategy using the best available scientific techniques to inform a political decision regarding the stakeholders' differing political values and perspectives on risk, relativity, and reward.

The Dimensions of Space Nuclear Power Policy

Considering the political feasibility of SNP, an agency such as NASA or DOD must face a three-dimensional transscientific program feasibility space. The objective is to find a policy option that balances the risks and rewards of SNP-enabled missions with the relative merits of the alternative power sources.

Figure 9 graphically depicts the dimensions of the problem by first considering the relative merits of SNP in comparison to such policy alternatives as solar energy along the vertical axis. The other two dimensions of policy analysis are measures of the risks (second axis) and rewards (third axis) of the mission.

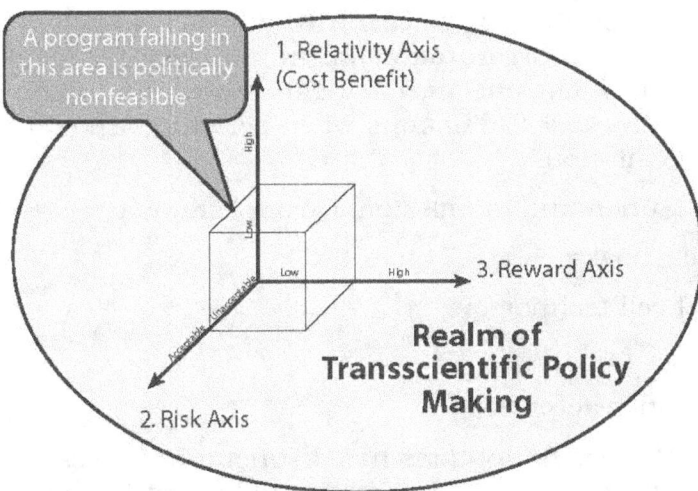

Figure 9. SNP program feasibility space.

Relativity (Axis 1)

Near the origin of the model, the benefits of SNP relative to the alternatives are low, the rewards are low, and risk is high. This creates an area where SNP is politically infeasible if planners seek tacit or explicit public permission, as the authors have argued. Moving away from the origin, the relative benefit of SNP increases with respect to the alternatives. Likewise, the other two axes depict increasing benefit and decreasing risk. In this outer area, SNP may be a politically feasible policy option.

This three-dimensional model can be applied to any program that NASA or DOD might contemplate, and using this model to reach a decision should be relatively straightforward. However, the transscientific nature of SNP complicates this process. The entire region of the feasibility space sits within a transscientific context. Because of this, the decision process is not one of simply determining the risk, relativity, and reward through a typical cost-benefit analysis. The transscientific nature of the problem means that there are technical, social, and political considerations that must be dealt with along all three axes of the model. It also means that there is no sharp boundary between feasible and nonfeasible space. The reader must apply these considerations to each axis.

The relativity axis measures the merit of each course of action. More specifically, it deals with the various power source options that an organization might consider for space missions. From a current and near future technical perspective, there are five discrete options when selecting a power source for space missions:

- no launch and no mission accomplishment,
- solar power,
- fuel cell technology,
- RTGs, or
- fission-reactor SNP.[3]

Each of these five options has distinct advantages and limitations. Obviously, the decision not to launch is a null option with no mission completed. This decision is particularly profound if it results because SNP was the only practical power source for a given mission. The consequence of the null policy option in this case is that it is a de facto political decision to cease practicing the science that explicitly defines it as a transscientific policy option.[4]

Where SNP is a viable option, its main advantages are high power, independent of distance from the Sun, and long life. The other three space power sources make technical tradeoffs between available power levels and operational life, with solar power restricted in operating range from the Sun. Cost is usually the driving factor when all other considerations are equal, but technical parameters such as energy density will be more important considerations for an interplanetary probe such as JIMO, where high energy, low mass, and high reliability will be required.

From a social perspective, the relativity involves tradeoffs between SNP-enabled missions and other programs that merit public support and consideration. For example, the primary mission objective of JIMO is to advance the life sciences. Observations from NASA's Galileo probe gave strong indications that Jupiter's moons (Callisto, Ganymede, and particularly Europa) have oceans underneath a sheath of surface ice. Astronomers believe these oceans to have more saltwater than the Earth and may have the prerequisites for the formation of extraterrestrial life. Life forms that draw energy from numerous different chemical processes on these moons could sidestep the need for sunlight or photosynthesis, on which most terrestrial life relies. JIMO proposes to explore the moons robotically for Jovian life forms and transmit the findings back to Earth. Certainly, this mission would be an historic boon to science. However, from the social perspective, launching JIMO must present neither a real nor a perceived unacceptable level of risk to the public. From a political perspective, the costs of JIMO must be acceptable to a democratic society and consonant with other national priorities.

Risk (Axis 2)

The second axis of the model depicts the real and perceived risk. From a technical perspective, this risk generally involves an analysis of the probability of failure of the system and the associated consequences of that failure. The transscientific nature of the problem limits the quantification of technical risk for SNP. Therefore, once technical risk analysis has been completed, the issues will quickly become political as much or more than they are technical. However, it is important to realize that the transscience aspects of the problem must eventually be reconciled with the rigorous requirements of science.

The fact that scientific knowledge is not value free will further complicate the issue. The social context of the researchers influences the issues.[5] Even experts suffer from cognitive biases, and transscientific issues impose significant limitations on risk assessments. Both technical and social analyses have quantitative and qualitative aspects. Technical risk analysis is not purely empirical, and the social is not as normative an enterprise as many would have it seem. Unfortunately, a robust framework from which to assess the junction of technical and social perceptions of risk is not available. Therefore, the transscientific nature of SNP makes the problem political more so than technical. Recognizing this fact will be critical in the formulation of policy.

In any case from the perspective of the public policy maker, risk is generally more than just a technical issue. Such other considerations as public understanding of the science and analytical methodologies and the public's risk tolerance play a critical role in the policy debate. Contrasting the political cultures engaged in the controversies over SNP makes it clear that political actors have great difficulty in understanding the fears and objections of others.[6] The process of discovery and intelligent risk mitigation through political action will be as intellectually challenging as the technical formulation of the options that pure science originally provided. In the end, a political decision maker will also have to decide how much political capital he is willing to risk to advance any program. In the case of SNP, this decision lies with the president. The calculation will be different for a NASA mission than for a national security issue. However, once the level of risk has been determined, the policy analysis is still not complete. The amount of risk that is acceptable—technically, socially, and politically—must also relate to the potential reward the program or mission might achieve.

Reward (Axis 3)

For NASA missions, the rewards generally center on basic science and a better understanding of the universe. For DOD, the rewards might include accomplishing a desired military mission, adding new capabilities, or maintaining a significant military advantage over potential adversaries. From a social perspective, the public must be convinced that the rewards are sufficient to merit support of the program. The reward must have recognized social value. Any SNP-enabled system or mission will be expensive. The public will not be willing to commit the required resources if they perceive that the reward is not sufficient to justify the cost.

The political acceptability of SNP for military missions will also depend upon SNP being separate from space weaponization issues. Current policy is that outer space will not be weaponized, and the United States has long-standing treaty obligations prohibiting weapons in space. Nevertheless, opponents of DOD argue that SNP is a prelude to the weaponization of space, and it is a fact that SNP is a potent enabling technology. The concerned public may also associate SNP with space weaponization. Therefore, the political feasibility of SNP depends upon severing any perceived linkages with space weaponization. Unlinking weaponization and SNP politically is perhaps the most important factor regarding public perception of SNP. Not including space weapons, some of the potential rewards of SNP are listed below:

- JIMO mission to the moons of Jupiter to search for life;

- space-based, SNP-enabled hyperspectral sensors to track terrorist and WMD proliferation threats (a hybrid ISR mission);

- space-based, SNP-enabled high-bandwidth communications for both military and civilian use; and

- emergent uses not currently envisioned.

Working in the three-dimensional realm of relativity, risk, and reward is a necessary but not sufficient condition for program decisions involving SNP. One has only to look to the history of failed SNP programs. For SNP to have any hope of success, it will first need to have a compelling mission requirement. The program will also have to navigate the murky waters of transscience and deal with a skeptical public. Going forward with the above model will require an effective engagement strategy, one that goes far beyond informing the public about mission plans.

In the landscape of strategy options, one methodology appears to hold significant promise. The methodology suggested here is the values-focused decision strategy introduced earlier.

Essentially, the values-focused decision strategy seeks a reasoned stakeholder consensus, which starts by recognizing the interested public as a valid stakeholder in the SNP debate, not just a passive victim of policy. Increasingly, the public will not accept being policy victims to big science. The overall aim of such an engagement strategy should be to bring to light such agreed ends as specifically enunciated and agreed goals of recognized public value from space science or national security with means, and estimates of the likely and tolerable risks surrounding those means, negotiated as part of the process. This may be unpalatable by conservatives in the policy-making fraternity, but the overall result should be ongoing program viability if such an engagement strategy is professionally pursued.

From the perspective of DOD, it might appear that there is little need or merit in pursuing a values-focused decision strategy. Clearly in the case of a national emergency or significant perceived threat, DOD can rely on the authoritarian reflex to gain permission to pursue SNP-enabled missions. As discussed in the previous chapter, an opportunity for this existed in the aftermath of 9/11. However, it is more likely that DOD will have to take a more measured approach to the development of programs using SNP. Furthermore, this process will necessarily occur over several presidential administrations, perhaps as many as five, so sustained support for the program will be necessary. Whether the general public will be involved in the values-focused process for national security systems is not clear. However, the Congress, acting on behalf of their constituents, certainly will be.

It seems clear, then, that both NASA and DOD could benefit from an engagement strategy that considers all stakeholders. In fact, NASA will have to pursue such a strategy all the time if it is to have any hope of achieving success with programs such as JIMO. DOD would be wise to employ an ongoing engagement strategy as well. First, the time horizon for development of SNP-enabled systems is quite long, perhaps 10–20 years. It is doubtful that permission resulting from a crisis situation can be sustained over that length of time. In addition, the benefits of a sustained engagement strategy would be enhanced should a national security imperative arise from a crisis. While the values-focused decision strategy is only one approach, experiments using this process indicate that it improves the chances of program success. The following section describes how the strategy applies to SNP.

A Values-Focused Decision Strategy for Space Nuclear Power

Values-focused decision strategies focus on how policy options are created and nuanced by the stakeholders, in contrast to the more common analytical techniques of dispute resolution.[7] This difference is important when working in transscientific endeavors such as SNP, where scientific techniques and analytical methods lack the political permissions necessary to empirically evaluate the policy alternatives. In fact, one advantage of a values-focused decision strategy is that it has the potential to minimize the most debilitating aspects of a transscience problem.

A values-focused decision strategy has five steps.[8] These are listed below and discussed in more detail in the succeeding sections.

- Carefully define the decision to be made.

- Identify what matters in the context of an impending decision in the form of the stakeholders' objectives.

- Create a set of appealing and purposeful alternatives.

- Employ the best available technical information to characterize the consequences of the alternatives, including the uncertainty associated with them.
- Conduct an in-depth evaluation of the alternatives by addressing the tradeoffs they entail.

Stakeholders play a major role in the formulation of values-focused decisions, and who these stakeholders are will depend on the mission. As discussed in the previous chapter, the primary kinds of missions for SNP involve space science, presumably under the direction of NASA; national security, under the purview of DOD; and possibly commercial applications, which could include a large number of international actors. Table 3 shows some of the stakeholders that are likely to be involved in the decision process for various missions.

Table 3. Potential Stakeholders in a Values-Focused Decision Strategy

Potential Stakeholder	NASA Scientific Missions	DOD Missions	Civilian Communications Missions
NASA	X		
Department of Energy	X	X	X
DOD		X	X
Congress	X	X	X
National Reconnaissance Office		X	
NSA		X	
NOAA	X		
Federation of American Scientists	X	X	X
UN		X	X
"In-Community" International Actors	X	X	X
Environmental Groups, e.g., Nukewatch or FCPJ	X	X	X

This table is not absolute or exhaustive. For any given program, the stakeholders may change, and some that might otherwise not be included may come into play. However, identifying a viable set of stakeholders will be critical to the decision strategy.

One of the more important aspects of the values-focused decision strategy is that it is, and is publicly perceived to be, participatory and democratic. Therefore, the process helps provide insights for the public into the reasoning behind transscientific policy decisions. It also creates a transparent process for all stakeholders and builds communication channels back to the public. As well, the strategy allows the stakeholders to gain insight to public concerns, and so develop responses to remedy difficulties.

Experience has demonstrated that a values-focused decision strategy can work. Participants in trials with *Cassini* as a test case expressed higher overall support for and satisfaction with *Cassini* than did those who worked under an expert-driven regime.[9] Done well, a values-focused decision strategy should explicitly address public concerns about permission, trust, liability, and risk.

Step 1. Defining the Decision

To define the decision, the proposing agency first must establish the mission requirements and the technological facts. Once the empirical data are on the table, the stakeholders compare the relative costs and risks of each feasible course of action. In this case, three major types of space missions that may seek to use nuclear power sources are military, scientific, or commercial. As described earlier, there are only five space power alternatives for these missions, and the analysis of the objective costs and technical feasibility of each option is straightforward.

SNP provides high power levels for long periods of time. All of the other power sources make significant technical trade-offs between available power and operational life. Given these scientific facts, the question is not whether we should use SNP, but rather how to balance mission capabilities, risks, and political rewards. There may well be a time in the future when SNP can be an alternative to solar or chemical energy. In the current political environment, SNP should be reserved for the missions that truly warrant the tradeoffs required and missions that cannot be deferred after considering the risks and rewards.

Step 2. Identify What Matters to the Stakeholders: Feasibility and Risk

The second step in the values-focused decision strategy is to identify what matters to stakeholders. This step-by-step approach helps to maintain the distinction between empirical facts and the interpretation of data. Values-focused decision strategies facilitate policy making by respecting legitimate differences in scientific and political judgment. By focusing on value judgments in this step, the strategy helps stakeholders to articulate their political objectives instead of becoming mired in the far more contentious activity of interpreting scientific data. Clearly stating the political objectives is a vital part of the strategy because politics, not science, defines specifically where the significant differences are and ultimately what matters in the debate.

Perhaps this appears as an intellectual defect from a scientific perspective, but political limitations born from the positions of the stakeholders turn out to be crucial to the development of public policy options that science cannot fully inform. By focusing attention on the political aspects of the problem as opposed to purely scientific aspects, the stakeholders can investigate implications of policy in crucial areas and in detail that would otherwise be impracticable for the scientific community to pursue alone. This is extraordinarily important, because transscientific policy is the articulation of the political acceptability of risks versus the perceived benefits of the policy options that science provides but cannot verify.

The proposing agency will consider SNP as a policy option when it appears to be the only technologically feasible way to accomplish the mission. Although the costs and technical feasibility are generally straightforward in the first step of this strategy, the risks and potential consequences of choosing a particular course of action are open to interpretation and value judgments by the stakeholders. Therefore, the interpretation of science informs the various risk and benefit assessments in transscience.

Considering the *Cassini* program as an example, proponents came from a small cross section of society. They tended to be highly logical technocrats imbued with enthusiasm for the romance of space exploration. That enthusiasm can be compelling, depending on the audience, and it colors the debate. The transcripts of public engagements between NASA and *Cassini* protestors indicate that the NASA representatives tended to push the scientific logic and analysis of their position, with the romance of space exploration serving to underpin their argument. On the other side of the debate, the FCPJ presented counterarguments to appeal to the fears of the public with regard to technology and to emphasize the absolute worst possible negative consequences extrapolated from scientific data. The opposing parties often talked past one another, and rarely did the different positions intersect such that they could initiate a negotiation.

The FCPJ and NASA were in a polarized and adversarial political battle in 1997 with little or no room for compromise or useful discussion of trade-offs. Yet, transscientific issues, especially when the science involves fissile materials, require a thoughtful understanding of the consequences that science informs but does not verify. To create dialogue between the stakeholders, the values-focused decision strategy could identify the key issues and potential area of compromise.

In a social experiment conducted by Arvai, researchers evaluated the differences between a values-focused decision strategy and expert-driven decision processes.[10] Arvai's study found that a values-focused strategy increased acceptance of a policy decision—either positive or negative—regarding RTGs on *Cassini* by roughly 25 percent for both men and women.[11] In the experiment, stakeholders subjectively evaluated overall acceptance on a seven-point scale. Approval rating increased from an average of 3.8 to 5.1 for women and 4.1 to 5.3 for men. The sample sizes indicated results were statistically significant to the 0.01 probability level for both men and women whether they were considering the risks or rewards of RTGs on *Cassini*. In the same study, the values-focused decision strategy improved the policy process without causing politics to interfere with the underlying science. The openness and democratic legitimacy of the values-focused strategy permitted the stakeholders to respect differing interpretations of science and therefore accept different policy outcomes.

Another example of a significant issue is the linkage between SNP and the weaponization of space. Although NASA generally makes a strong case against this linkage, they still must deal with it. And, there can be little doubt that the linkage to weaponization will immediately be made for any DOD system that proposes to use SNP. How program proponents address this issue with the relevant stakeholders will have a major impact on the success of the policy process.

Step 3. Create Appealing and Purposeful Alternatives

The third step in the values-focused decision strategy is to create appealing and purposeful alternatives. In considering space power alternatives, the technical limitations are quite clear. Nuclear power sources (both RTG and fission) require special consideration because they have political considerations that solar and chemical technologies do not. The overall political acceptability must balance the mission benefits with a technical risk assessment of alternative power sources.

Agencies such as DOD may soon be proposing nuclear power to support national security objectives. SNP is particularly attractive to DOD because the department already uses both nuclear and space technology and has resident expertise. Therefore, the costs of entry are considerably lower for DOD than for other government agencies. DOD also might tend to prefer nuclear power sources because they have a much higher energy density than the alternatives and are considerably more reliable. Here is a short list of the missions that DOD may propose that may or may not require SNP.

Military missions that may require nuclear power:

- space-focused radar,
- laser communications,
- electric or ionic propulsion,
- space-based data processing,
- midcourse discrimination,
- advanced meteorology,
- battlefield illumination, and

- directed energy station keeping.

Military missions not likely to require nuclear power:

- GPS or navigation,
- continuous optical or UV reconnaissance,
- space weather,
- meteorology, and
- communications.

Planners generally cannot scale military space missions up or down to move from one power option to another. For example, space-based radar sensors require a continuous power source of greater than 100 kWe to provide space sensors with the ability to generate sufficient radiant energy to map the Earth's surface as well as to transmit data with the reliability required under wartime conditions. Therefore, with currently available technologies, space-based ground-mapping radar will require SNP, and such alternatives as fuel cells or solar energy will not suffice for mission requirements. Nevertheless, DOD's opponents contend that missions might be accomplished without space resources. This results in choosing the option of not launching rather than risking a military mission with fissile material in orbit.

This may seem like a dilemma where appealing alternatives are not available to either camp. However, it may be possible to reduce the power requirements in space to make solar or fuel cell technology more attractive and then augment the space mission with airborne assets such as the Air Force's Joint Stars or the U-2. The values-focused decision strategy encourages the stakeholders to examine these options in depth.

For scientific exploration of space, it is possible to scale required power levels up or down to a much greater extent than with military missions. The inflexible aspect of scientific mission requirements is usually flight time. For example, devices requiring low levels of power will suffice for interplanetary explorations. However, an interplanetary expedition is necessarily a long-duration operation and will require a highly reliable power source that can function for months or even years.

Therefore, scientific missions tend to prefer one of three power options. Solar power is long lasting but limited by low solar flux beyond Earth orbit. Also, solar panels are large and quite fragile, which diminishes their reliability and puts the mission at risk. This risk is important when considering the costs of space expeditions and the national prestige that may be on the line in the event of mission failure.

RTGs are also limited in the amount of power they can provide. However, they are extraordinarily reliable and can function continuously for as long as 70 years without maintenance. This has made RTGs the preferred power source for deep space exploration thus far, despite the fact that there was considerable political opposition to launching RTGs containing Plutonium 238 on the *Cassini* mission in 1997.

When SNP is considered for scientific space missions, it must be considered only for the few missions that require considerable power for long periods of time. SNP will also be necessary when long-duration power is required and solar panels are not sufficiently reliable for the mission. The scientific missions that may meet those requirements are listed below.

Potential scientific endeavors requiring nuclear power:

- deep-space missions,
- interplanetary robotic exploration,
- extended human lunar missions,
- human exploration of Mars,
- propulsion, and
- advanced meteorology.

The potential commercial applications of SNP are currently limited to hypothetical scenarios only. Due to the transscientific nature of SNP, it is highly unlikely that commercial interests would be the first to use nuclear technologies in space. It is more likely that commercial ventures would piggyback on a military or scientific enterprise. Only after SNP has passed from the transscience arena into the realm of routine public policy will we see space nuclear reactors dedicated to commercial enterprises. Some of those future enterprises are listed here.

Potential commercial endeavors potentially requiring nuclear power:

- space-based manufacturing,
- laser communications,
- high-power and broadband communications, and
- directed energy power transmission.

The potential rewards of SNP used for commercial purposes are primarily financial; however, it would be extraordinarily difficult to justify public risks for the sake of private investment interests in the present political environment. Perhaps commercial uses of SNP will be justifiable sometime in the future after the risks have been mitigated by previous experimentation and SNP has moved beyond the realm of transscientific policy making. In the interim, commercial use of SNP will likely be limited to joint enterprises with the military or scientific communities.

When stakeholders perceive the technical options are exhausted, they will undertake more normative work to politically reformulate the policy options in terms more favorable to their biases and interests. Stakeholders must attempt to resolve residual ambiguities concerning risks and reward to reach an acceptable consensus. The stakeholders refine consequence descriptions and perform political risk analysis of problems to which scientific technique had previously only drawn attention.

Step 4. Employ Best Available Technical Information to Characterize Consequences and Uncertainties

By articulating what matters and working together to create appealing alternatives, the stakeholders tend to focus on and attempt to increase the accuracy and scope of science that they interpret to be particularly revealing of either the political risks or the benefits of a proposed program depending upon their preferred policy option. The stakeholders focus on scientific facts that compare directly with political predictions from their preferred policy options. Stakeholders such as Nukewatch, the Federation of American Scientists, or the FCPJ are willing to invest great effort and ingenuity to bring science and their preferred policy options into closer empirical agreement.

Ultimately, when the stakeholders' analysis and interpretation of science cannot serve their particular interests, the stakeholders must endeavor to discover revealing new facts or else adjust their interests and begin to accept some compromise. Even so, some established stakeholders acting as social risk amplifiers occasionally refuse to consider new scientific analysis and show little tolerance for those who do.[12] For example, Nukewatch activist Bruce Gagnon has protested that scientific exploration missions to the Mars area are "dangerous step in the expansion of nuclear technology into space" because the spacecraft uses RTGs.[13] Mr. Gagnon insists that these missions are dual use with the military, despite the fact that there is no conceivable military mission beyond Earth orbit, and the RTGs used are 30-year-old technology with no applicability to weapons development.

Nevertheless, science can eventually extend the stakeholders' experience base to facts that will be particularly persuasive. This step in the decision strategy is an effort to move policy goals forward using the scientific tools that are available, including the untested, incomplete, and unverifiable hypotheses that normal science can provide without incurring unacceptable political risks. According to Steven Aftergood of the Project on Government Secrecy within the Federation of American Scientists:

> There are those in the broad general public who aren't concerned about safety. There are others who can't be convinced that space nuclear power and propulsion are sufficiently safe. But for everyone else, a program that is open, accountable, and responsive to public inquiries is most likely to be acceptable. As a practical matter, this means acknowledging that space nuclear power, like spaceflight in general, is not "safe" in any absolute sense. The issue rather, is the value of the mission as well as the adequacy of the steps that are taken to minimize potential hazards.[14]

This dialogue is beneficial to both scientific interests and the political debate. If risks and consequences are characterized with intellectual integrity, then this grounds the political debate with sound scientific judgment and analysis. The political context likewise proscribes ethical scientists from performing experimentation or research that may cause unacceptable public risks and should have political oversight. At this juncture the policy maker simultaneously has useful political and scientific leverage on the problem. This dual leverage is necessary for transscientific policy to advance science and minimize public risks. The transparency of the process results in a more open-ended policy formulation model that respects the democratic process.

Step 5. In-depth Evaluation of Alternatives Addressing Tradeoffs

The process of discovery and intelligent risk mitigation through political action can often be as intellectually challenging as the original technical formulation of the options that pure science originally provided. The previous steps of the values-focused decision strategy require reasonable effort to call forth new data and political options. The art of policy process addresses the concerns of both sides of the political debate in a scientifically sound way.

At this final step in the decision strategy, the options are evaluated based on their political merit as informed by the scientific judgments carefully formulated in the previous steps. The JIMO mission provides an example. The expedition will need nuclear power unless there are extraordinary advances in solar or chemical energy sources, and the authors do not foresee such advances. Waiting for them could be a matter of decades. If JIMO is to proceed, even by 2012, it must utilize SNP. Therefore, the principal public risk with JIMO involves getting fissile material into orbit safely.

In this step of the values-focused decision strategy, the stakeholders evaluate the alternatives using best professional practices and scientific techniques. Although judgments will vary, stakeholders specify the alternatives in scope and detail that would not be possible otherwise. Where differences in judgment exist, those judgments are specific and based upon common entering arguments for scientific evaluation. Values-focused strategies considering stakeholders create options and permit policy decisions to be accepted by the stakeholders as compared to more didactic methods of dispute resolution.[15] This is important when working in such transscientific endeavors as SNP, where scientific techniques and analytical methods lack the political permissions necessary to evaluate the policy alternatives scientifically.

How to Know if the Process is Working

One fundamental question about the process remains: How will one know that the values-focused engagement strategy is working? Determining measures of effectiveness is an important challenge, but it is necessary to ensure that the strategy is meeting objectives. Because of the transscientific nature of the problem, purely empirical measures are necessary but not sufficient to judge how effective the strategy is in attaining its goals. Therefore, subjective evaluations must supplement empirical measures of merit. Examples are

- assessment of proponents' and opponents' political will,
- opportunity costs of not accomplishing a scientific mission,
- additional risk to national security of not accomplishing a military mission, and
- potential environmental damage from launch failure.

These measures of effectiveness will factor into the policy makers' advice to the political decision makers regarding SNP. Although there is a great deal of technical information that can be quantitatively determined, the qualitative assessments of the risks and rewards are the politically contentious issues. Because the risks and rewards cannot be determined precisely or strictly quantified, the interpretation of science by the stakeholders should explain the concerns and why they matter politically.

Operationalizing a Values-Focused Decision Strategy

The United States will soon be at a decision point concerning SNP. The fact that a nuclear technology in space is a transscientific issue complicates the balance between SNP's promise and the public's concern about public risk. Policy makers cannot ignore the scientific uncertainty of the consequences of a nuclear payload breaking up in the Earth's atmosphere. Technical experts may argue over the ramifications of an accident, and program opponents and proponents may use their own interpretations of science to further their policy goals. In the meantime, the public in the middle remains quietly confused and certainly uncomfortable with the consequences program opponents describe. The transscientific doubt about the SNP places Congress and the executive branch in the difficult position of making a decision about SNP programs without a complete understanding of either the risks or the consequences.

Because transscientific decisions are ultimately a matter of political judgment, the risks of SNP should be considered along with reward and relativity. These three policy dimensions are the foundation upon which a values-focused decision strategy aids policy formulation. Ultimately, the goal of a values-focused decision strategy is to manage the transscientific uncertainties while protecting the intellectual integrity of the underlying science that originally generated the policy option. A values-focused decision strategy outlines a democratically legitimate and scientifically sound rigorous method to assist policy makers in considering SNP as a transscientific policy problem. Given the high stakes surrounding SNP, an effective engagement strategy is critical to the success of any future SNP-enabled missions.

Notes

1. Victoria Friedensen, "Protest Space: A Study of Technological, Choice, Perception of Risk, and Space Exploration" (master's thesis, Virginia Polytechnic Institute and State University, 11 Oct 1999).

2. R. L. Keeney, *Values-Focused Thinking: A Path to Creative Decision Making* (Cambridge, Mass.: Harvard University Press, 1992).

3. American Institute of Aeronautics and Astronautics (AIAA), Aerospace Power Systems Technical Committee. "Space Nuclear Power: Key to Outer Solar System Exploration," an AIAA position paper, Reston, Va.: AIAA, 1995.

4. Thomas S. Kuhn, *The Structure of Scientific Revolutions* (Chicago: University of Chicago Press, 1962), 151.

5. Sheldon Krimsky and Dominic Golding, *Social Theories of Risk* (Westport, Conn.: Praeger, 1992), 356.

6. Carlo C. Jaeger et al., *Risk, Uncertainty and Rational Action* (London: Earthscan, 2001), 186.

7. Gerald W. Cormick and Alana Knaster, "Oil and Fishing Industries Negotiate: Mediation and Scientific Issues," *Environment* 28, no. 10 (Dec 1986): 6–16.

8. Joseph L. Arvai, Tim L. McDaniels, and Robin S. Gregory, "Exploring a Structured Decision Approach as a Means of Fostering Participatory Space Policy Making at NASA," *Space Policy* 18, no. 3 (Aug 2002): 221–31.

9. Jeremy Hsu, *Project Prometheus: A Paradigm Shift in Risk Communication*, sciencepolicy.colorado.edu/gccs/2003/student_work/deliverables/jeremy_hsu_project_prometheus.pdf (accessed 8 Mar 2004), 8.

10. Arvai, "Exploring a Structured Decision Approach."

11. For more specific statistical data see ibid., 228.

12. As an example, see the following court case where the plaintiff repeatedly tries to suppress environmental impact statements revised specifically to answer the plaintiff's original scientific objections. *Florida Coalition for Peace and Justice v. George Herbert Walker Bush*, in *Lexis*, vol. 12003.

13. David Leonard, "NASA's Nuclear Prometheus Project Viewed as Major Paradigm Shift," http://www.space.com/businesstechnology/technology/nuclear_power_030117.html (accessed 26 Mar 2004).

14. Ibid.

15. Cormick, "Oil and Fishing Industries Negotiate," 6–16.

Chapter 5

Conclusions

I began by trying to quantify technical risks, thinking that if they were "put into perspective" through comparison with familiar risks, we could better judge their social acceptability. I am ashamed now of my naiveté, although I have the excuse that this was more than twenty years ago, while some people are still doing it today.

—Harry Otway, PhD, 1992

Nuclear technologies were the first instance of a scientific problem that entered the political realm as transscience. If mishandled, the negative consequences of nuclear experimentation would have been momentous, if not catastrophic. A public increasingly aware of the risk has slowly withdrawn political permission for conducting most nuclear science. Since the Three Mile Island nuclear accident in 1979, the United States has not built nor ordered any new nuclear power plants. The issues of waste disposal continue to be contentious. Like most things nuclear, space exploration has significant risk that complicates the political problems of SNP. Therefore, we believe that the most certain way to cultivate the unscientific public's trust is to simultaneously improve the underlying nuclear and space technologies while engaging the public politically in an open and democratically transparent way.

The authors also believe scientific technique will eventually be able to fully inform SNP policy. Presently, however, scientific techniques and analytical methods lack the political permission necessary to properly evaluate the risk and reward of SNP as an alternative source of power in space. In the meantime, the authors believe that a public policy decision that involves SNP should be made with a values-focused decision strategy that informs the political dimensions of risk, relativity, and reward.

This paper addressed the question, "What mechanism would improve the political feasibility of a nuclear power program for US space operations?" The authors' inquiries have highlighted the fact that the answer is contextual and falls squarely in the realm of political judgment as opposed to scientific analysis. Moreover, the difficulties of making these political judgments are intrinsic to the transscientific nature of SNP. Policy makers have incomplete empirical or analytic data to draw from when making decisions. The interpretation of available data can vary widely from one set of experts to the next, clouding judgment about social value and political risk in choosing whether to implement such a transscientific program as SNP.

Unlike pure science, transscientific policy must harmonize the rigors of scientific judgment with the imperatives of politics. Empirical analysis is a necessary but not a sufficient tool for solving transscientific policy problems. That conclusion and the fact that some stakeholders have not assimilated it have caused significant problems in engaging the public with respect to SNP. There is a resulting discomfort about SNP, even in politically moderate circles, because of the dearth of empirically verifiable data regarding the political risks and rewards. Although the rewards of SNP are potentially profound, they are sacrificed because of publicly perceived but unverifiable risk.

Although SNP is technically feasible, it is not currently politically acceptable to the public because of the perceived risk. The present situation contrasts sharply with the 1950s, when SNP enjoyed tacit public support and was even touted in the popular press. This change seems to be because the unscientific public has become increasingly aware of partisan risk assessments presented by political activists opposed to all forms of nuclear power, including SNP. However, the public is still largely unaware of the transscientific uncertainties regarding the assessment of both risk and reward.

Organizations such as the FCPJ or Nukewatch have made the maximum plausible assertions regarding the possible risk, while NASA and/or DOD have made minimum plausible assertions about this same risk. These boundary positions have done little to advance either science or reasoned political discourse. Neither group can be sure of its position with respect to SNP because of the transscientific doubt surrounding the risk and potential consequences of an accident on the launch pad or in Earth's atmosphere.

The issues surrounding transscientific risk assessment for SNP are extraordinarily complex and multifaceted. Therefore, an engagement strategy is required that is democratically legitimate and transparent. A strategy must take the issue to the people, or at least to their representatives. A decision strategy that addresses such difficult scientific issues must also protect the intellectual integrity of science from undue political influence.

To assist in addressing stakeholder concerns, the authors have presented a model that policy makers considering SNP can use to ensure they have thoroughly evaluated the positions of the various stakeholders while simultaneously respecting the rigorous requirements of sound scientific judgment. The authors' first recommendation is that potential risk should be considered along with reward and relativity, much as with any other political issue. However, in the case of SNP, these three considerations exist within a transscientific context. The transscientific nature of the problem implies that sound scientific analysis is a necessary but not a sufficient prerequisite to achieving a politically acceptable solution. The political interpretation of scientific data and value judgments will determine whether SNP is feasible or not.

NASA has been carrying the torch for SNP and will continue to do so until DOD establishes a definite and politically acceptable mission imperative. Therefore, NASA's engagement strategy must establish the initial issues around SNP and space science that are germane. The DOD should learn from NASA's experience, even if society's national security permissions arise from different concerns than do those of space science. Although the mission objectives would be different, the political dimensions of risk, relativity, and reward, shrouded by transscience, would be the same for both agencies. We also believe that DOD should consider an ongoing program that would at least maintain the state of the art in SNP technology, as well as explore mission scenarios where SNP might be desirable or necessary.

There are several reasons for doing this, but at least three stand out. First, the knowledge base for developing SNP is rapidly aging. The last major program to be considered was terminated in 1993 and many of the scientists in the field have either retired or moved on to other things. Second, it is likely that future DOD space operations will require high power, the kind that only nuclear reactors appear to be able to provide. Given the long lead-time to develop such systems, an ongoing program makes sense. Third, SNP is not a system enabler that can be developed and deployed overnight. Should a future national security situation arise where SNP would play a critical role, an ongoing program is much more likely to be available in the short term than one started from scratch.

The authors' next recommendation is to use a values-focused decision strategy to inform the dimensions of the transscience feasibility space as discussed in chapter 4. A values-focused strategy presents an alternative to the present, highly polarized political framework. In the values-focused alternative, both sides agree to seek a proposition based on shared values.[1] By determining what matters to the stakeholders in terms of informed value judgments, concerns regarding risk and reward are shared early and embedded in policy formulation from the outset. In essence, this values-focused strategy seeks a reasoned stakeholder consensus that respects sound scientific judgments on both sides of the debate. The strategy also recognizes the interested public as a valid stakeholder in the SNP debate.

The values-focused decision strategy allows the stakeholders to inform the unscientific public about their interpretation of scientific information regarding both the risks and rewards of SNP. These value judgments about the interpretation of scientific data are folded into the political decision process and publicly reevaluated using the best available scientific techniques to characterize the consequences of a particular course of action. This analysis, in turn, informs the classical political dimensions of risk, relative merit, and reward. Two of the more important aspects of the values-focused decision strategy are that the strategy is participatory and democratic. Therefore, the process helps to provide insights for the public into the reasoning behind transscientific policy decisions. This process creates a transparency for all the stakeholders and builds communication channels between the concerned public and the policy makers. NASA is in an excellent position to test the strategy much as was done for *Cassini* project.[2] This test should focus on Project Prometheus and the proposed JIMO mission. If successful, the values-focused decision strategy could then be employed with the relevant stakeholders.

For more than 50 years, the United States has explored the potential of nuclear power in a variety of space-focused applications. There have been numerous technical challenges; however, most of the technical issues have now been overcome. The public is interested in space science but is also sensitive to the political risks, the relative merit of alternatives, and the potential rewards. Politically aligned and activated, even a small part of the public would pose pressure that policy makers could not ignore, and such pressure may determine the feasibility of SNP systems going forward.

Will the day arrive when there are nuclear reactors flying in space? NASA certainly hopes so, and DOD will probably wish it to be so in the not too distant future. The ever-increasing demands for power in space seem to suggest that SNP is inevitable. No matter what the outcome, we believe informed stakeholders making decisions based upon shared values, while still respecting legitimate differences in scientific judgment, is a far better policy process than the divisive and uncompromising situation that presently exists.

Summary of recommendations:

- Modern society is risk adverse, especially so regarding nuclear technologies.
- Carte blanche permission for the US government to develop nuclear technologies has been withdrawn.
- For SNP to be feasible, there must be a compelling mission requirement and a reasonable level of popular political support.

- The values-focused decision strategy should be employed by both NASA and DOD SNP missions.

- NASA should carry SNP public policy forward with a focus on deep space exploration—a mission that requires SNP technology.

- Department of Defense also has a strong case for SNP arising from emerging security concerns and should leverage NASA's experience.

Notes

1. Joseph L. Arvai, Tim L. McDaniels, and Robin S. Gregory, "Exploring a Structured Decision Approach as a Means of Fostering Participatory Space Policy Making at NASA," *Space Policy* 18, no.3 (Aug 2002): 221–31.
2. Ibid.

Appendix A

The History of Space Nuclear Power

> *The utilization of atomic power in outer space offers the greatest opportunity for the United States to create a climate of world good will through leadership in peaceful space exploration. This matter could be as important as the hydrogen bomb debate in the early 1950s or the Manhattan project decision in 1942. Few things deserve more serious reflection, for we can now be missing a great opportunity while sowing the seeds of another prestige disaster.*
>
> —Sen. Clinton P. Anderson
> Chairman of the Joint Committee on Atomic Energy
> US Senate, 1960

Early Developments

Before embarking on a discussion of the history of SNP, this appendix considers the political context in which the early programs began. The Cold War was underway and tension between the United States and Soviet Union was increasing. The first use of atomic power had been the two bombs dropped on Japan by the United States at the end of the Second World War in 1945. The Soviet Union conducted its first atomic test in 1949 and by 1953 both superpowers had detonated hydrogen bombs. Both countries were rapidly developing large stockpiles of nuclear weapons. Against this backdrop, Pres. Dwight D. Eisenhower initiated a bold program to pursue nuclear power for peaceful purposes. In a speech before the United Nations General Assembly on 8 December 1953 he described a new initiative called Atoms for Peace. President Eisenhower explained his policy in the following quote:

> The United States knows that peaceful power from atomic energy is no dream of the future. That capability, already proved, is here now—today. Who can doubt, if the entire body of the world's scientists and engineers had adequate amounts of fissionable material with which to

test and develop their ideas, that this capability would rapidly be transformed into universal, efficient and economic usage?[1]

This speech indicates that even with the development of nuclear weapons there was a sincere desire to see other applications of nuclear energy come into being. It is also an indication that the leaders of American society, if not the entire world, were encouraging support for the development of nuclear technologies. In some sense, there was probably no more opportune time for the development of SNP.

The Nuclear Navy Program

Even before the Atoms for Peace initiative, the United States was in the process of pursing nuclear power. In 1948 the Navy began the development of a nuclear reactor for submarine propulsion. The Navy effort is worthy of analysis because it highlights several important contextual issues which combined to make it successful. In contrast, the political factors that made the Navy program successful did not transfer directly to SNP.

The Navy program was the first major nuclear power effort after World War II. The Cold War was beginning and the political imperative to establish worldwide US military superiority had high priority. At the time, a submarine's capability to operate submerged was limited. They were basically surface ships with the ability to submerge for periods of only 30 to 40 minutes. While under water their maneuverability was also extremely limited. Their poor subsurface capability imposed severe tactical limitations on their employment and compromised their most important attribute, stealth.

Hence, a main motivation for developing nuclear power for submarines was to obtain the capability to operate for extended periods below the surface of the ocean. The nuclear submarine program achieved a significant milestone on 17 January 1955 with the *Nautilus*.[2] The *Nautilus* was the world's first nuclear-powered naval vessel. Now submarines could stay underwater for extended periods of time with a reliable power supply that could provide propulsive as well as onboard electrical power. The success of the *Nautilus* spawned future generations of ever more capable nuclear-powered submarines and initiated profound changes to the fundamentals of naval warfare, deterrence, covert operations, and intelligence gathering.

Underlying the success of the Navy program were several important factors. First, the Navy had a compelling mission requirement that no available alternative enabling technology could provide. Second, nuclear propulsion produced a substantial performance increase for submarines that improved their capability, employability, and versatility. Third, the Navy had the advantage of economy of scale in that many submarines would ultimately be produced, making the development of purpose-built nuclear reactors cost effective.

The Navy also benefited from the tacit, carte blanche, political permissions to proceed with nuclear technology that the Cold War provided. The public knew little about nuclear power and national security was adequate justification to proceed with nuclear power on warships. In fact, at the time there was no requirement for regulatory approval outside of DOD. In addition, the political effectiveness of then Capt Hyman Rickover should not be underestimated.[3] His constant direction, push, and oversight contributed to the success of the program.

Finally, the Navy program appears to have had wide acceptance by American society to proceed. Even the entertainment industry was politically supportive. Walt Disney Productions created a graphic for the *Nautilus*. The political conditions reflected by the image stand in stark contrast with those of today. It is almost impossible to imagine today an entertainment company signing on to promote a politically incorrect and contentious technology such as SNP. Walt Disney Productions also produced an animated film entitled *Our Friend the Atom* in 1957 that was intended to promote atomic technology.

Early Space Reactor Programs

Proposals for SNP began to appear in the late 1940s. Despite the earlier successes of the Navy program, few policy makers seem to have seriously considered the uses of nuclear power for space applications. Rather, public attention was turned to the development of commercial nuclear power reactors. Work was begun on the Shippingport nuclear reactor located on the banks of the Ohio River in Pennsylvania. This reactor was to be the world's first commercial nuclear electric power generator. By all accounts the Shippingport program progressed successfully, and the nuclear core of the reactor was inserted on 4 October 1957.[4]

Four October 1957 is significant for another event, the launch of sputnik. The political shock of the sputnik launch to the United States was momentous and galvanized US political will into pursuing the space race against the Soviet Union. The country placed increasing emphasis on scientific education and began to invest significant time and resources in space programs. Around the time of sputnik, scientists and engineers who had been working on the Shippingport reactor were already thinking about where the next new challenge for nuclear power might come from. They got their answer as they gazed upward to see the small Soviet satellite pass overhead and wondered about the role for nuclear power in space. This crucial question, even at the earliest stages of SNP, is indicative of the major problems that plagued the SNP program. What are the mission requirements and what alternative courses of action are available to policy makers? The following quote from Glenn Seaborg, then the chairman of the Atomic Energy Commission summarizes the scope of what the United States wanted to accomplish, "What we are attempting to make is a flyable compact reactor, not bigger than an office desk, which will produce the power of the Hoover Dam from a cold start in a matter of minutes."[5]

Early Space Propulsion Reactor Programs

Initial studies on nuclear rocket propulsion had begun as far back as 1947.[6] The technical focus was on nuclear rockets for missile propulsion. The first perceived science mission requirement for a nuclear-propelled rocket was a mission to either Mars or the Moon. However, unlike the Navy program, no mission was in the works and the initial development occurred before President Kennedy's speech that launched the program to put a man on the Moon.

This lack of a mission requirement that could be met with another technology was probably the most significant issue that would hobble the SNP program for decades to come. Over the years, the United States considered several nuclear rocket propulsion systems, but one that stands out for its technological promise is the Nuclear Engine for Rocket Vehicle Applications (NERVA) program. NERVA became a major program effort in early 1961. Developed over several years, the NERVA program was an outstanding technological success. Planners intended NERVA for a mission to Mars because the NERVA rocket had several payload advantages. The most significant was the takeoff weight envisioned for the Mars mission. Chemical rockets would have produced an orbital mass three to five times greater depending on planetary alignment.[7]

A flight test for NERVA was scheduled for 1973 but the program was abruptly discontinued in late 1972. The program was cancelled due to a lack of funding and the lack of definite plans for a Mars mission. As the program was being pursued, various political issues related to nuclear materials began to cause delays. The reactor thrust plume was slightly radioactive, although the total radioactivity was not considered a significant hazard. The greatest effect of the radioactive plume came when the aboveground test ban treaty was signed in 1963. After the treaty became effective, all aboveground testing was prohibited—even for peaceful purposes such as NERVA. In the coming years, political constraints, like the aboveground test ban treaty, would continue to haunt the development of SNP.

Early Space Power Reactor Programs

At the same time that the United States was pursuing nuclear-powered rockets, the Eisenhower administration made a decision in 1960 to develop a nuclear reactor system for electrical power in space. That system, designated the Space Nuclear Auxiliary Power (SNAP)-10A and developed jointly by the Atomic Energy Commission and the Air Force, was flown in orbit in 1965. The design of the reactor was technically elegant. Several safety procedures were put into place to limit the risk of a nuclear accident and exposure to radioactive materials. First, the reactor was launched in *cold state* meaning the reactor would not actually be turned on until it was safely established in its proper orbit. Second, even during operation, its design prevented the reactor from going out of control. Finally, the reactor was in a very high orbit above the Earth, approximately 1,296 kilometers (over 800 miles). The reactor functioned flawlessly for 42 days and then shut down after a voltage regulator failure. It remains in orbit today and is not expected to reenter the Earth's atmosphere for several hundred years.[8]

SNAP-10A validated many technologies and operational procedures and expectations for the future of SNP soared. An advertisement for SNAP-10A appeared in *Newsweek* magazine in the 21 June 1965 issue. This is significant as it points to the kind of political permission that existed at that time. This advertisement is not unlike one that would appear today for a sports car or modern appliances. The possible uses for the reactor were quite varied and the company that produced it, North American Aviation, appeared optimistic about the future of SNP.

Despite the technological successes of the SNAP-10A, no additional systems were produced and the program apparently ended without follow-on development. The reasons for this seem lost, but one likely explanation is the lack of a compelling mission that demanded the capabilities only a SNP system could provide. The government focused on the lunar program, and it appears to have taken precedence over many other programs. It is unfortunate that SNP fell off the national priority list because when it appeared again in the 1980s the political conditions were quite different. By the 1980s the public's tolerance for nuclear technology was much diminished from that of the 1950s and 1960s.

The Soviet Union's Space Nuclear Program

At the same time that the United States was developing both nuclear propulsion and power systems, the Soviet Union was also working hard on SNP for military applications. From 1967 to 1988 the Soviet Union launched 35 nuclear reactor systems.[9] The primary purpose for these reactors was to provide power for Radar Ocean Reconnaissance Satellites (RORSAT). Soviet radar technology was limited at the time, and to obtain a useful radar signal the satellites had to be flown at low orbital altitudes. This altitude was too low for solar panels to be used due to their high atmospheric drag, hence the need for SNP. This highlights a major difference between the Soviet program and that of the United States. Unlike the United States, the Soviet Union did have a compelling mission requirement that could not be satisfied by an alternative technology.

The Russian reactor, called the Topaz 1, was first unveiled at the 1964 Third UN Conference on the Peaceful Uses of Atomic Energy. These reactors were used until the late 1980s, well into the timeframe where worldwide political activism concerning nuclear power was beginning to have significant policy impact elsewhere in the world. Unlike the United States, the Soviets had no need to consider the internal political aspects of using SNP. Furthermore, given the highly sensitive nature of these systems, it is likely the Soviet public did not even know they existed.

Significantly, the Soviet RORSAT program suffered five failures, three of which resulted in unplanned reentries of fissile material. In April 1973, a Soviet RORSAT mission launch failure resulted in the return of the power source in the Pacific Ocean, North of Japan. Air sampling by United States planes detected radioactive material as would be expected from the Topaz 1 reactor. On 24 January 1978 a Cosmos satellite carrying a Topaz 1 nuclear reactor reentered the atmosphere over the Northwest Territories of Canada. According to the Soviets, the reactor was designed to burn upon reentry. Nevertheless, a significant amount of radioactive debris was found in an area of approximately 100,000 square kilometers. This reentry was a major international incident and certainly would have been much worse had the breakup occurred over a population center such as Montreal or Toronto. Another reactor, this time from *Cosmos 1402*, reentered the atmosphere in February 1983 after the payload was not boosted into a high enough orbit. Fortunately this reactor fell harmlessly into the South Atlantic Ocean although a trail of radioactive material was left behind in the atmosphere.[10]

The Soviet Union apparently stopped using space nuclear reactors in the late 1980s. Clearly, concern about the previous accidents was a factor. In fact, the 1978 accident accelerated the eventual decline of the RORSAT program. Furthermore, the Soviet Union had turned its attention to submarines as a means to track US aircraft carriers. In addition, Soviet radar technology had advanced sufficiently that lower orbit operations were not required to accomplish the desired mission. It is possible that the use of SNP would have ended sooner than it did. However, it was only after Mikhail Gorbachev came into power that new and different ways of thinking were allowed to be explored.[11]

These Soviet failures have had important policy implications for the US program. They highlight the potential risks of spaceflight and the serious consequences when nuclear materials are involved, particularly when the environmental consequences of an accident cannot be fully assessed due to the transscientific nature of this problem. By the 1980s the general public's concern over nuclear issues made the prospects of a new SNP program politically infeasible even though technological feasibility was well established.

Into the 1990s

Despite the fact that early SNP programs never advanced beyond the test and evaluation phase, development continued into the 1990s. After a nearly 20-year hiatus, interest in the United States was rekindled with the Strategic Defense Initiative (SDI). Some information from this era remains classified, but two major programs emerged. The first of these was Timberwind, a nuclear rocket propulsion system. The other was SP-100, a nuclear reactor that was intended to provide primary power for several space-based components of SDI. When initially proposed, these two programs appeared to satisfy the long-missing compelling mission requirement for SNP. SDI was ambitious, complex, and would require electrical power levels never before envisioned. However, as was the case for much of SDI, the requirement disappeared with the end of the Cold War in 1989, and so did Timberwind and SP-100.

Timberwind

The Timberwind rocket was a heavy lift vehicle that would launch SDI components into orbit. In many ways, the development of this program mirrors that of the NERVA rocket. Timberwind began in 1982 and it is estimated that more than

$800 million was spent on the program. The program was largely declassified in 1992 and renamed the Space Nuclear Thermal Propulsion (SNTP) program. The Clinton administration cancelled the program in 1993 when it terminated most nuclear programs.

Once information about Timberwind became available several scientists and other activists raised serious concerns. Quite vocal and ominous sounding, their pronouncements were significant in contrast to the near total silence surrounding the NERVA program. One example comes from the late Dr. Henry Kendall, then chairman of the Union of Concerned Scientists and a Nobel Laureate. He is quoted as saying: "The needle just goes up on the end of the [danger] scale and stays there. Such a rocket would release a stream of radiation as it flew and if it broke up, 'you've got radioactive material spraying all over the place' . . . the risks are extremely great."[12]

If nothing else, this comment and others like it indicate that the tacit political permission previously granted by society was under serious challenge. By the 1980s, people were fearful of nuclear technologies. Appeal and amplification of that fear, if not heard by everyone in the society, certainly were ringing in the ears of policy makers, particularly as organized social protest began to emerge.

SP-100

The SP-100 program started in 1983. It was the first major reactor program to undergo serious development since SNAP-10A. SP-100 was also an important program for SDI. The 100 in the name comes from the plan to have the reactor produce 100 kW of electric power. It was intended to operate for seven years of full power over a 10-year life cycle. The SP-100 design allowed for power outputs ranging from 10 kW to one megawatt.[13]

The SP-100 program appears to have been plagued by problems almost from the beginning. First, SP-100 was complex and pushed the edge of technology in materials, fuel, and energy conversion. Program participants included the Defense Advanced Research Projects Agency (DARPA), DOD, Department of Energy (DOE) and NASA. As the system evolved, the time to completion was pushed farther out. This introduced significant program risk, as support was necessary over several sessions of Congress and three different administrations.[14]

Surprisingly, there appears to have been little organized protest against the development of SP-100. This can partially be explained by the classified nature of the program. In addition, the major protests were over the SDI in general, not particular systems. Finally, SP-100 was cancelled in 1993.

The *Cassini* Mission

In 1997 NASA launched a space probe named *Cassini*.[15] Dubbed the "Rolls Royce" of space missions, it is one of the largest and most ambitious missions ever undertaken. The *Cassini* probe explored Saturn and conducted numerous experiments on the planet, its rings, and satellites. The cost of this mission was about $3.4 billion. The *Cassini* probe arrived at Saturn in July 2004.

Those unfamiliar with the *Cassini* mission might wonder why it is significant in the political debate about nuclear power in space. More to the point, *Cassini* is not powered by a nuclear reactor, the main focus of this paper. However, it is largely powered by nuclear energy, that energy coming from an array of RTGs. RTGs were discussed in chapter 1. What makes them particularly significant for *Cassini* is the quantity of plutonium 238 being carried on the probe; that is, 72 pounds. This is the largest amount ever launched into space. The launch of this amount of a highly toxic substance caused significant protest.

Protestors challenged NASA to verify the safety of the program. Under the umbrella of the FCPJ, a number of protest events were held and legal action was taken against the mission. Concern centered on two aspects of *Cassini*. First was the threat of explosion on launch with the resulting possibility of contamination in the launch area. However, the larger concern centered on the fact that the *Cassini* probe was going to make a close approach to the Earth in 1999. The fear was that if the probe reentered the Earth's atmosphere it would burn up and large amounts of plutonium would then enter the ecosystem. For the protestors, this risk was unacceptable and doomsday proclamations were made about the future of humanity. The close approach occurred in 1999 without incident.

The highly toxic nature of plutonium is what frightens people the most. Even extremely small amounts lodging in the body, particularly the lungs, are thought to cause lung cancer. Fear of a plutonium release is not without precedent. In 1964 a US navigational satellite failed to reach its intended orbit. The RTG power source disintegrated into the atmosphere as it was designed. This incident released 2.1 pounds of Pu-238 into the atmosphere, tripling the worldwide inventory. There has been considerable debate about the potential health impacts of radioactive materials released into the atmosphere by these accidents. Some have argued that significant increases in worldwide cancer can be directly attributed to the additional plutonium in the atmosphere.[16] However, there is no way to scientifically confirm this. While it is true that Pu-238 is now found in the bones of all human beings as a result of the 1964 reentry, it is also true that the average life span of humans has been steadily increasing.

Here again we see the transscience nature of SNP. Questions about the potential risks of nuclear materials release can never be answered by experiment and testing. No rational actor would be willing to explode a functioning space nuclear reactor on launch or allow it to burn up in the atmosphere to see what would happen. We are left to make the best technical analysis and estimates and then make a political calculation about the risk, relativity, and reward of any mission involving SNP.

The purpose of this paper is not to explore *Cassini* in depth; this has already been done elsewhere. The reader is invited to review the excellent work of Victoria P. Friedensen, which chronicles the details of the protest movement and its consequences.[17] Another discussion of the issues surrounding *Cassini* can be found in an online forum conducted by the *Newshour* program on PBS.[18] To gain a perspective on the opposition's point of view, the book by Karl Grossman, *The Wrong Stuff*, is the most extensive source.[19]

Despite the risks, the *Cassini* mission went as planned. However, it is clear that political permission from society was being challenged in significant ways. Had the FCPJ been successful in delaying the launch of *Cassini*, it is possible the mission might not have ever flown. Unlike the early days of the

Atoms for Peace initiative, many in the society saw nuclear energy more as Atoms for Death. This was especially true for space missions where the direct benefit to society was called into question.

Contemporary Space Nuclear Power

After the termination of Timberwind and SP-100 in 1993, there was no major program to develop SNP. In fact, in 2001 DOD's *Space Technology Guide* dropped any reference to SNP. Defense officials at the time were quoted as saying: "In the STG, the Congress asked for an investment strategy for space technology. Given the severely constrained funding available for space technology development, funds for nuclear power devices would not make the priority cut. Even if we could produce them economically, the mission costs would be unaffordable because of the measures necessary for security."[20]

There are no major technical hurdles that would prevent further development and deployment of space systems using nuclear reactors for power or propulsion. Technical feasibility has already been well established and demonstrated with SNAP-10A in 1965. Given sufficient time and resources, there is little doubt that the technology can be made safe to twenty-first century standards. In fact, space-based reactors have one safety advantage over RTGs. The reactors can be launched in a nonoperating mode, and turned on only after the proper orbit is achieved. One could in fact argue that there is really no restriction on developing SNP to the same level of safety and low operating risk that the airline industry now enjoys. Nevertheless, it is true that the environmental consequences of an SNP accident on launch or in the atmosphere remain transscientific. That, in the context of today's political climate, will mean that the use of SNP will remain a contested political decision.

The political feasibility of SNP will be tested with the announcement of NASA's Project Prometheus. NASA has decided to embark on a new deep-space mission to explore the icy moons of Jupiter. Among the many challenges facing this mission is the requirement to use a nuclear-reactor-based propulsion system. See Appendix B for a description of Prometheus from a technical and scientific perspective.

Once again, a program to develop SNP has been initiated. If history is any indicator, this project faces significant challenge. NASA no longer has the luxury of developing technologies in a political vacuum. Several activists' groups are certain to oppose the project. NASA will be forced to engage the public to gain acceptance for the program. Given the long and troubled history of SNP, it is hard to imagine this will be an easy undertaking.

Notes

1. Dwight D. Eisenhower, Atoms for Peace, speech given to United Nations General Assembly, 8 Dec 1953.

2. John W. Simpson, *Nuclear Power from Undersea to Outer Space* (La Grange Park, Ill.: American Nuclear Society, 1995), 65.

3. Ibid., 35.

4. Ibid., 109.

5. Ibid., 115–20.

6. NERVA, http://www.astronautix.com/project/nerva.htm (accessed 24 Mar 2004).

7. Simpson, *Nuclear Power from Undersea to Outer Space*, 151.

8. Joseph A. Angelo and David Buden, *Space Nuclear Power* (Malabar, Fla.: Orbit Book Company, 1985), 165–68.

9. Steven Aftergood, "Background on Space Nuclear Power," *Science and Global Security* 1 (1989): 97.

10. Ibid., 98–100.

11. Prof. Roald Sagdeev, interviewed by Lt Col Jim Downey at Eisenhower Institute, 24 Feb 2004.

12. "Global Network—Plutonium in Space (Again!)," http://www.globenet.free-online.co.uk/articles/morenukesinspace.htm (accessed 24 Mar 2004).

13. Aftergood, "Background on Space Nuclear Power," 100–103.

14. Mitch Nicklovitch, PhD, interviewed by Lt Col Jim Downey, Oct 2003.

15. The reader is invited to review the excellent work of Victoria P. Friedensen, which chronicles the details of the protest movement and its consequences. Another discussion of the issues surrounding *Cassini* can be found in an online forum conducted by the *Newshour* program on PBS. To gain a perspective on the opposition's point of view the book by Karl Grossman, *The Wrong Stuff*, is the most extensive source. These references are in the bibliography.

16. Karl Grossman, *The Wrong Stuff: The Space Program's Nuclear Threat to Our Planet* (Monroe, Maine: Common Courage Press, 1997).

17. Victoria Friedensen, "Protest Space: A Study of Technological Choice, Perception of Risk, and Space Exploration" (master's thesis, Virginia Polytechnic Institute and State University, 11 Oct 1999).

18. Online Newshour Forum, "The Cassini Mission—21 October 1997," http://www.pbs.org/newshour/forum/october97/cassini.html (accessed 24 Mar 2004).

19. Grossman, *The Wrong Stuff*.

20. "DOD's Flirtation with Nuclear-Powered Satellites Ends, Analyst Says," http://www.fas.org/sgp/news/2001/03/iaf030201.html (accessed 24 Mar 2004).

Appendix B

Project Prometheus—
Frequently Asked Questions—
December 2003

This appendix is reproduced from NASA, "Project Prometheus, Frequently Asked Questions," December 2003 http://www.jpl.nasa.gov/jimo/Schiff_FAQ_03_pdf2.pdf (accessed 11 March 2004).

What is Project Prometheus?

NASA's Mission to understand and protect our home planet, to explore the Universe and search for life, and to inspire the next generation of explorers requires that we make strategic investments in technologies that will transform our capability to explore the Solar System and beyond. Within the Space Science Enterprise, we are developing the tools, insights, and abilities necessary to answer some of humanity's most profound questions: How did the Universe begin and evolve? How did we get here? Where are we going? Are we alone?

In Greek mythology, Prometheus was the wisest of the Titans who gave the gift of fire to humanity. The word "Prometheus" is synonymous with "forethought," an idea that embodies NASA's hope to establish new tools for expanding our exploration capabilities.

NASA believes that, in the field of space exploration, answering these questions translates to constantly striving to develop innovative scientific instruments and more effective ways to safely power, propel, and maneuver spacecraft, as we explore the worlds beyond our current reach. Achievement of this ambitious vision requires a bold approach to the next generation of solar system exploration missions, including revolutionary improvements in energy generation and use in space.

Project Prometheus, the Nuclear Systems Program, is making strategic investments in near- and long-term nuclear electric power and propulsion technologies to maintain our current space science capabilities and that would enable space exploration missions and scientific returns never before achievable. In addition to developing the next generation of radioisotope power systems, the predecessors of which have been used for over 30 years to power space science missions, Project Prometheus would develop and demonstrate the safe and reliable operation of a nuclear reactor-powered spacecraft on a long-duration space science mission. Toward this end, the proposed Jupiter Icy Moons Orbiter (JIMO) mission has been identified as the first space science mission that would incorporate these new revolutionary technologies.

What is the Jupiter Icy Moons Orbiter?

Project Prometheus is developing a proposed space science mission, the Jupiter Icy Moons Orbiter, that would enable detailed scientific investigation and data return from the icy moons of Jupiter - Callisto, Ganymede and Europa - that may have three ingredients considered essential for life: water, energy and organic material. Making use of nuclear fission power and electric propulsion, the mission would involve one spacecraft orbiting, at close range and for long durations (months at a time), these three planet-sized moons. The spacecraft would orbit each of these moons for extensive investigations of their makeup, history, and potential for sustaining life.

In addition to enabling an entirely new class of scientific investigations, the mission would also demonstrate the safe and reliable use of a space nuclear reactor in deep space for long-duration space exploration. The amount of power available from a nuclear reactor—potentially hundreds of times greater than that available to current interplanetary spacecraft—would enable delivery of larger payloads with vastly more capable instruments and faster data transmission back to Earth than such missions as Voyager, Galileo, and Cassini. In addition, because extremely fuel-efficient electric thrusters would propel the spacecraft, mission planners could make course adjustments throughout the mission in response to real-time discoveries.

How will the Jupiter Icy Moons Orbiter get to Jupiter?

After being launched from Earth by a traditional chemical rocket, the spacecraft would rely on an electric propulsion system, e.g., a system expelling electrically charged particles called ions from its engines to generate thrust. Powered by a small nuclear reactor, the electric propulsion system would propel the spacecraft to the Jovian system and then insert the spacecraft into orbit around, successively, each of Jupiter's three icy moons. In 1998, NASA's Deep Space 1 mission successfully demonstrated the use of ion propulsion for interplanetary travel.

Why do you need to use a nuclear reactor to get to Jupiter?

The large quantities of power generated by the compact nuclear reactor (about the size of a 5-gallon bucket) enable a variety of advanced mission capabilities, and therefore increased scientific return, not possible with conventional power systems.

To start, access to high levels of continuous power enables full-time maneuverability of the spacecraft. After being launched from Earth, using conventional chemical rockets, the spacecraft would use fuel-efficient electric thrusters to propel it to Jupiter. Once in the Jovian system, the engines would propel the spacecraft to each moon where controlled, close-range orbits would provide ideal conditions for science observations. Moreover, the maneuverability afforded by nuclear power and electric propulsion would enable mission scientists to alter mission plans based on real-time discoveries. Such maneuvers are not possible with current chemical propulsion systems, which consume the bulk of their propellant during departure from Earth. Once outside Earth's gravitational influence, such conventionally powered spacecraft coast to their destination, making very limited adjustments to their trajectories using relatively inefficient chemical combustion for propulsion.

Electric propulsion should also enable delivery of significantly heavier, and in most cases larger, payloads to destinations throughout the solar system and beyond. Because the electric propulsion system would consume fuel very efficiently, it could be used throughout the mission to gradually accelerate the spacecraft and its unprecedented scientific payload to high velocities.

Either in transit or at its final destination, the reactor could power high-capability, active science instruments never before used beyond Earth orbit. Another benefit is that space scientists could operate these instruments simultaneously rather than cycling them as is currently the practice due to limitations in the power available on present-day exploration missions.

Finally, the reactor would power the spacecraft's communications equipment, which would transmit the voluminous science data acquired by these instruments back to Earth in quantities and speeds never before possible.

Why are Jupiter's icy moons a priority (i.e., what is the scientific justification)? What are the science goals?

Exploring the Universe and searching for life are central to NASA's mission, and Jupiter's large icy moons appear to have three ingredients considered essential for life: water, energy and the necessary chemical elements. NASA's Galileo spacecraft found evidence for subsurface oceans on these three moons of Jupiter—a finding that ranks among the major scientific discoveries of the Space Age.

The National Research Council (NRC) completed a report last year, based on input from the planetary science community, that prioritized potential flight missions for exploring the solar system. It ranked an "Europa geophysical explorer" mission as its top priority for a "flagship" mission, based on the Galileo data suggesting a liquid ocean under Europa's ice crust. The Jupiter Icy Moons Orbiter mission would build upon and exceed the NRC's recommendation by not only conducting in-depth investigations of Europa, but because of the propulsion capabilities of the spacecraft, it would also examine Callisto and Ganymede, providing comparisons key to understanding all three.

The Jupiter Icy Moons Orbiter mission has four major science goals:

1. Determine the interior structures of the icy moons of Jupiter in relation to the formation and history of the Jupiter system;
2. Determine the evolution and present state of the Galilean satellite surfaces and subsurfaces, and the processes affecting them;
3. Determine how the components of the Jovian system operate and interact, leading to the diverse and possibly habitable environments of the icy moons;
4. Determine the habitability of Europa and the other icy moons of Jupiter.

How much fuel will be used in the reactor?

The amount of fuel needed for the proposed reactor would depend on the final reactor design. However, based on a reference power level of 100 kilowatts (electric power), the reactor would be quite small: the entire reactor core could fit within a 5-gallon drum. Each of the reactor design concepts currently being considered could have from 100 to 150 kilograms or approximately 220 to 330 pounds of uranium fuel.

How hazardous would it be if there were an accident in space and the reactor explodes?

The reactor will be designed with multiple safety features that will prevent uncontrolled, sustained nuclear fission (which could disassemble the reactor) before, during or after launch of the spacecraft. Moreover, the space reactor would be designed to remain intact over a broad range of ground and in-space accidents.

How often will a mission similar to the Jupiter Icy Moons Orbiter be mounted?

It is planned that the Jupiter Icy Moons Orbiter mission would be the first of many scientific missions enabled by nuclear electric power and propulsion. Specifics regarding any future missions would be dependent on NASA's exploration requirements, developed in consultation with the scientific community. It is expected that the technologies developed by Project Prometheus could support future space exploration missions, including human exploration of space.

What will be the operations cost for the JIMO mission and how long will those costs continue?

The operational cost will be highly dependent on the method chosen to implement the Jupiter Icy Moons Orbiter mission. When the initial mission studies are completed in FY05, NASA expects to be in a better position to provide accurate and complete project life cycle cost estimates, including operational costs.

What confidence do we have that the systems developed for the JIMO mission will work?

NASA is very confident in its ability to design and build systems that will meet all mission requirements and be ready to launch by the early part of the next decade, assuming that the required funding is received. The technical hurdles, while significant, do not require major breakthroughs but can be managed with a focused and consistent engineering effort by the nation's R&D community.

Where will we get the enriched uranium for the JIMO reactor?

The uranium-235 for the fission reactor would come from current federal government stocks owned and managed by the Department of Energy.

What is the total cost of this mission? Why is it so expensive?

NASA's funding estimates reflected in the President's FY04 budget run out through FY08 includes roughly $3 billion for Project Prometheus, with just over $2 billion of that directed toward the technology development for the Jupiter Icy Moons Orbiter and similar missions. NASA expects to be able to provide more accurate and complete project life cycle cost estimates when initial mission studies are completed in FY05.

The Jupiter Icy Moons Orbiter mission would be the most capable deep space science mission ever launched by NASA, with revolutionary new capabilities enabled by the power available from its space nuclear reactor. NASA views Project Prometheus and the JIMO mission as strategic investments necessary to expand our capabilities to effectively support our mission of exploring the universe and searching for life.

What scientific payload could justify the expense of such a mission?

Presently, because of power limitations, outer solar system exploration missions have been limited in their science capabilities and, therefore, their science return. Project Prometheus is NASA's strategic investment in technologies that could provide the science community with the energy supply necessary, practically anywhere in the solar system, to dramatically increase science opportunities and the quality of science conducted throughout a mission.

Because of the power available from a space nuclear reactor, the spacecraft would be able to carry instruments with capabilities far beyond those flown in previous outer solar system missions, including high-power, active instruments as well as instruments of much greater precision and resolution that would generate orders of magnitude more data than current missions with significantly smaller power sources. Examples of such instruments are high-power radars that could penetrate deep into the subsurface of the three moons (10s of kilometers) in search of liquid water, more capable cameras and spectrometers with greater resolution (200 colors vs. 7 colors or less than 100 meters per pixel vs. 100 km per pixel) to map nearly the entire surface of each moon, and instruments that use lasers to measure the topography of, or to illuminate, extraterrestrial surfaces. Moreover, as opposed to current missions where instruments are cycled on and off, the nuclear-powered spacecraft would have the capability to power all its science instruments, if desirable, simultaneously.

When assessing the science potential of a mission, one must look not only at the scientific payload, but also at how these instruments can be used throughout the mission. Nuclear electric power and propulsion technologies are being studied because they have the potential to enable close-range observations of multiple destinations for extended periods of time in a single mission; to adjust mission objectives in response to real-time discoveries, and to transmit huge amounts of data back to Earth.

Specific science instruments have yet to be identified for the proposed JIMO mission. To facilitate this process, NASA is working with the science community via a Science Definition Team, to identify specific science objectives for the mission and the measurements necessary to support these objectives. In addition, NASA has begun a new program dedicated to developing the new high capability instruments that would be possible on the potential JIMO mission. Final determination of science instruments would be carried out through an Announcement of Opportunity (AO), which would provide the science community an opportunity to formally propose specific instruments and measurements.

How will you ensure planetary protection at Europa?

NASA's Office of Space Science will work with the NASA Planetary Protection Advisory Committee to develop guidelines for planetary protection requirements for Europa. For example, this mission would be designed to reduce the probability that the spacecraft would strike Europa at any stage of the mission.

Why can't this mission be done with solar electric propulsion and solar sails?

Solar arrays would have to be far too large to produce the electrical power required to operate the electric propulsion system and scientific instruments for the Jupiter Icy Moons Orbiter mission. The Sun's energy at Jupiter is less than 1/25th of its level at Earth, which make this type of mission virtually impossible to perform with solar arrays, even taking into account expected improvements in solar array efficiency in the foreseeable future.

NASA is researching the ability of solar sails (not to be confused with large solar arrays) to enable low mass spacecraft to achieve large increases in velocity by using the pressure of sunlight to fill a lightweight sail, thereby "pushing" the spacecraft. Solar sails may one day be used to propel small spacecraft to the outer solar system, but presently their most effective use appears to be within a "zone" no more than twice the Earth's distance from the Sun. Therefore, solar sails are not a viable option for propelling a JIMO-like spacecraft to Jupiter, let alone maneuvering it around Jupiter's three icy moons.

What is the Department of Defense's role in this program? Is NASA really just a front for DOD in their desire for fission-powered space weapons?

While NASA maintains open lines of communication with various components of the federal government, DOD has no role in this program. Project Prometheus program requirements have and will continue to be established to meet NASA's science and technology needs for space exploration.

Meanwhile, NASA works very closely with the Department of Energy (DOE) to develop space science missions using nuclear power sources. As we expand such cooperation to include both radioisotope and nuclear reactor power systems, we will be calling upon more of DOE's experience base and technical infrastructure than that necessary for radioisotope work alone.

How do you plan to test the reactor to meet the schedule for this mission?

We are too early in the program definition phase to appropriately address this topic. Specific details of how the space reactor will be tested would be defined once a reactor type has been selected and the design, test, and manufacturing plans are worked out with the reactor developer.

How safe are radioisotope thermoelectric generators and reactors?

Safety is of the utmost importance and drives the overall design of radioisotope power systems and reactors, their applications, and the extensive testing, analysis and review that each system undergoes. Prior to any mission carrying nuclear material, NASA and the Department of Energy (which is responsible for development of any space nuclear systems for NASA) jointly conduct extensive safety reviews supported by safety testing and analysis. To date, NASA has safely developed, tested, and flown radioisotope power systems on 17 missions and the United States successfully launched a nuclear reactor into earth orbit in 1965. The Department of Energy and NASA place the highest priority on assuring the safe use of any nuclear power systems for space missions.

In addition to internal agency reviews for missions involving nuclear systems, an ad hoc Interagency Nuclear Safety Review Panel (INSRP) is established as part of the Presidential nuclear safety launch approval process to evaluate the safety analysis report prepared by the Department of Energy. Based upon recommendations by the Department of Energy and other agencies and the INSRP evaluation, NASA submits a request for nuclear safety launch approval to the White House Office of Science and Technology Policy (OSTP). The OSTP Director may make the decision or refer the matter to the President. In either case, the process for launch cannot proceed until nuclear safety launch approval has been granted.

Launch approval for United States space missions that use nuclear systems is based on careful consideration of the projected benefits and risks of the proposed mission. The analysis of potential consequences will be based on a detailed understanding of: a) the possible accident environments; b) the response of the nuclear system in those accident environments; c) modeling how any potential releases of nuclear material might be transported; d) estimates of potential public exposure and the consequences of those exposures.

What is the danger to the public from this project? (testing, launch, flight, re-entry?)

NASA's top priority is to ensure that this program and its missions can be implemented safely. Therefore, safety will be the primary driver in every aspect of the program, including spacecraft design, test, manufacture, and operation. All program activities will be conducted in a manner to reduce risk to levels as low as reasonably achievable. A hierarchy of safety objectives, requirements and engineering specifications will be established and followed during each phase of every mission.

To support these objectives, NASA will identify and mitigate risks as early in the system design process as possible and we will work continuously to ensure the safety of the public, workers, and the environment. NASA will provide opportunities for public review and comment throughout the life of the program.

What were NASA's previous failures with space nuclear systems? Have there been any failures by other organizations or nations?

None of the more than thirty radioisotope power systems and one reactor system flown by the United States has failed. Three missions using radioisotope power systems have been subject to mechanical failures or human errors resulting in early aborts of each mission. In each instance, the radioisotope power system performed in accordance with its design requirements.

The first such incident occurred during the launch of a Navy navigation satellite in 1964. The Navy satellite failed to achieve orbit and burned up on re-entry, which was in keeping with the safety design practice at that time. Subsequent radioisotope power systems were designed to remain intact on re-entry. In 1968, NASA aborted its Nimbus-B weather satellite two minutes after launch because human error had caused the rocket to veer off course. The Radioisotope Thermoelectric Generator (RTG) was retrieved intact from the Santa Barbara Channel off the coast of California. The fuel from that system was reused on a subsequent NASA mission. Lastly, in April 1970 an RTG survived the breakup of the Apollo 13 Lunar Module and went down intact in the 20,000 foot deep Tonga Trench.

In 1965, the United States successfully launched a small nuclear reactor into Earth orbit. The reactor operated safely until an electrical failure, unrelated to reactor operation, caused it to prematurely shut down. The spacecraft is now orbiting Earth at a distance that will ensure that it does not return until its radioactive fuel has been rendered harmless due to radioactive decay.

Open literature suggests that through 1988 the former Soviet Union launched just over 30 nuclear-powered spacecraft into Earth orbit for marine radar observations. In 1978 a Soviet space reactor re-entered the atmosphere and landed in Canada. Five years later, a Soviet reactor re-entered over the South Atlantic Ocean. In 1996, the Russian Mars 96 spacecraft carrying an RTG failed to reach Earth orbit on launch and fell into the eastern Pacific. It is believed that some of the debris may have fallen over South America.

Who is involved in the launch pad (safety) processes?

NASA's Kennedy Space Center (KSC) has the overall management and integration responsibilities for launch site ground processing operations. However, launching NASA spacecraft is a joint effort between KSC and the Air Force at Cape Canaveral Air Force Station and Patrick Air Force Base. The Air Force is always involved in launch activities through their management of range assets. For mission launches where the spacecraft incorporates a space nuclear power system, the Department of Energy provides on-site assistance in monitoring and risk assessment.

How many RTG launches might there be in the next 10 years?

At least two NASA missions within the next decade are considering the use of radioisotope power systems, but it is expected that others may also pursue this option. The New Horizons mission to Pluto and the Kuiper Belt is in development for a launch in 2006 with a radioisotope thermoelectric generator to supply electricity to the spacecraft. The Mars Science Laboratory, in development for a 2009 launch, is considering two new radioisotope power systems currently under development by NASA and the Department of Energy—the Multi-Mission Radioisotope Thermoelectric Generator and the Stirling Radioisotope Generator. Aside from these two missions, Project Prometheus is working closely with the space science community to identify missions that could take advantage of the unique capabilities enabled by a radioisotope power system.

Is the research really worth the cost?

NASA recognizes that Project Prometheus is a major investment of taxpayers' dollars for which there should be equally significant benefits. Although difficult to quantify in monetary terms, NASA firmly believes that the technologies developed through Project Prometheus have the potential to revolutionize our ability to explore and understand better the Universe and, ultimately, humankind's past, present, and future.

In the near term, this strategic investment in new technologies would enable an entirely new class of exploratory missions from which unprecedented science data would be returned. Propelled by extremely efficient electric thrusters, a nuclear reactor-powered spacecraft could observe multiple destinations at close range and if necessary, even modify mission objectives mid-mission based on real-time discoveries. Onboard, the spacecraft would carry science instruments that could peer into unknown worlds with more precision and clarity than ever before imaginable. In specific terms, the amount and quality of the data returned from the Jupiter Icy Moons Orbiter mission would dwarf that of any other robotic mission to the outer solar system. New ranges of science instruments, never used beyond Earth's orbit, would capture and return more data than the two Pioneer, two Voyager, Galileo, and Cassini missions combined.

Meanwhile, development of new radioisotope power systems delivering just over 100 watts of power would enable the Mars Science Laboratory to operate anywhere on the planet, regardless of the location of the sun, for months or years rather than days or weeks. Similar power sources could provide electricity for small-to-medium size space missions such as those proposed for New Horizons. Smaller power systems under consideration by NASA, from milliwatts to several watts, would provide mission planners a full complement of long-lived, rugged, reliable power sources. There will likely be significant technological benefits to areas outside of space exploration as well.

In the long term, the technologies developed in support of the Jupiter Icy Moons Orbiter mission could be evolved to the larger power and propulsion systems necessary to support human exploration beyond Earth orbit. Additionally, the knowledge and technologies developed through Project Prometheus-sponsored research and development will have broad applications throughout NASA, other parts of government, academia, and the private sector.

For over thirty years, NASA has relied on the same set of power and propulsion systems to explore the solar system and beyond. Project Prometheus is the investment necessary if NASA is to take a major step forward in our quest to explore our solar system and search for life, and through its groundbreaking technologies and missions would provide an inspiration to the next generation of students and explorers.

Appendix C

The Member Groups of the Florida Coalition for Peace and Justice or Stop Cassini! Campaign

This list is taken from Victoria Friedensen, "Protest Space: A Study of Technology, Choice, Perception of Risk and Space Exploration" (master's thesis, Virginia Polytechnic Institute and State University, 1999).

Action Network for Social Justice (Tampa)
Affirmation Lutheran Church (Boca Raton)
Alliance for Survival (Costa Mesa, CA)
Bangladesh Astronomical Society (Dhaka)
Brevardians for Peace and Justice
Broward Citizens for Peace and Justice
Cassini Redirection Society (British Columbia)
Catholic Diocese of Jacksonville, Office of Peace and Justice
Center Florida Council of Churches
Center for Advancement of Human Cooperation (Gainesville)
Central Florida Presbytery Polk County
Citizens for Peace and Justice
Citizen Soldier (New York, NY)
Coalition Freedom Coalition (Gainesville)
Community Action Network (Seattle, WA)
Crow Indian Landowners Assoc. (Montana)
Cuba Vive
Darmstadter Friedensforum (Darmstadt, Germany)
Delray Citizens for Social Responsibility
East Bay Peace Action (Berkeley, CA)
Environmental & Peace Education

Florida Southwest Peace Education
Global Peace Foundation (Mill Valley, CA)
Global Resource Action Center for the Environment
Glynn Environmental Coalition (Brunswick, GA)
Grandmothers for Peace (Elk Grove, CA)
Grandparents for Peace (St. Augustine)
Iowans for Nuclear Safety (Cherokee, Iowa)
Jacksonville Coalition for Peace and Justice
Jonah House (Baltimore, MD)
Kalamazoo Area Coalition for Peace and Justice
Leicester Campaign for Nuclear Disarmament
Mama Terra Romania (Bucharest)
Maryland Safe Energy Coalition (Baltimore)
Metanoia (Jacksonville)
North Palm Beach Unitarians (Deland)
No Nukes Action Project (Los Angeles, CA)
North Country Coalition for Justice and Peace
Orlando Friends Meeting
Patriots for Peace (Ft. Walton Beach)
Pax Christi (Florida)
Peace Action (Washington, DC)
Peace Links (New York, NY)
Peace Resource Center of San Diego
People's Action for Clean Energy (Canton, CT)
Phillip Berrigan (Baltimore, MD)
Plutonium Action, Hiroshima (Japan)
Presbytery of Tampa Bay Peace of Highlands County
Radiant Medicine Project (Kingman, KS)
Sisters of Mercy Social Justice Team (Brooklyn)
Solar Design Associates (Harvard, MA)
South Florida Peace Network
St. Margaret Mary Catholic Church (Winter Park)

Tallahassee Society of Friends
Tampa Bay Peace Education Program
The Nuclear Resister (Tucson, AZ)
Unitarian Friends Fellowship of Pineda
Unitarian Social Concerns Committee Gainesville
Unitarian Universalist Church of Sarasota
Unitarian Universalist Church of Tampa
Unitarians at Large (Del Ray Beach)
Unitarians Jacksonville
Unitarians Miami
Unitarians Unitarian Fellowship Vero Beach
Ursulines of Tildonk for Justice and Peace
United States-Vietnam Friendship Assoc. (San Francisco)
Vets for Peace (Gainesville)
Vets for Peace (Tallahassee)
Volunteers for Peace (Belmont, VT)
War & Peace Foundation (New York)
War Resisters League (Asheville, NC)
Westminster Presbyterian Church (Lakeland)
Women Strike for Peace (Washington, DC)
Women's International League for Peace and Freedom - Tampa
Women's International League for Peace and Freedom - Treasure Coast
Women's International League for Peace and Freedom - West Palm Beach Winter Park Friends Meeting
Women's International League for Peace and Freedom (Palm Beach County)

Selected Bibliography

Aftergood, Steven. "Background on Space Nuclear Power." *Science and Global Security* 1 (1989): 93–107.

American Institute of Aeronautics and Astronautics (AIAA). Aerospace Power Systems Technical Committee. "Space Nuclear Power: Key to Outer Solar System Exploration." An AIAA position paper. Reston, Va.: AIAA, 1995.

Angelo, Joseph A., and David Buden. *Space Nuclear Power*. Malabar, Fla.: Orbit Book Company, 1985.

Arvai, Joseph L.; Tim L. McDaniels; and Robin S. Gregory. "Exploring a Structured Decision Approach as a Means of Fostering Participatory Space Policy Making at NASA." *Space Policy* 18, no. 3 (Aug 2002): 221–31.

Beck, Ulrich. *The Risk Society: Towards a New Modernity*. Thousand Oaks, Calif.: Sage Publishing, 1992.

Butler, Amy. "DOD's Flirtation with Nuclear-Powered Satellites Ends, Analyst Says." *Inside the Air Force*, 2 Mar 2001. http://www.fas.org/sgp/news/2001/03/iaf030201.html (accessed 24 Mar 2004).

Cormick, Gerald W., and Alana Knaster. "Oil and Fishing Industries Negotiate: Mediation and Scientific Issues." *Environment* 28, no. 10 (Dec 1986): 6–15, 30.

Dahl, Robert A., *On Democracy*. New Haven, Conn.: Yale University Press, 1998.

Dahl, Robert A., and Edward R. Tufte. *Size and Democracy*. Stanford, Calif.: Stanford University Press, 1973.

Eisenhower, Dwight D. "Atoms for Peace." Speech. United Nations General Assembly, New York, 8 Dec 1953.

Federation of American Scientists. "Anti-Ballistic Missile Treaty." http://www.fas.org/nuke/control/abmt/.

Friedensen, Victoria. "Protest Space: A Study of Technological Choice, Perception of Risk, and Space Exploration." Master's thesis, Virginia Polytechnic Institute and State University, 11 Oct 1999.

Grossman, Karl. "Alternative Energy Meets Main Street." *New Age*, July/Aug 1999, 59.

———. "Plutonium in Space (Again!)." *Covert Action Quarterly*, no. 73 (Summer 2002). http://www.globenet.free-online.co.uk/articles/morenukesinspace.htm.

———. *The Wrong Stuff: The Space Program's Nuclear Threat to Our Planet.* Monroe, Maine: Common Courage Press, 1997.

———. To Wing Commander Anthony Forestier. E-mail, 9 Oct 2003.

Hsu, Jeremy. *Project Prometheus: A Paradigm Shift in Risk Communication.* sciencepolicy.colorado.edu/gccs/2003/student_work/deliverables/jeremy_hsu_project_prometheus.pdf (accessed 8 Mar 2004).

Jaeger, Carlo C. et al. *Risk, Uncertainty and Rational Action.* London: Earthscan, 2001.

Jasanoff, Sheila S. "Citizens at Risk: Cultures of Modernity in Europe and the U.S." *Science as Culture* 11, no. 3 (2002): 363–80.

Kammen, Daniel, and David Hassenzahl. *Should We Risk It? Exploring Technological Problem Solving.* Princeton, N.J.: Princeton University Press, 1999.

Keeney, R. L. *Values-Focused Thinking: A Path to Creative Decision Making.* Cambridge, Mass.: Harvard University Press, 1992.

King, David, and Zachary Karabell. *The Generation of Trust: How the United States Military Has Regained the Public's Confidence since Vietnam.* Washington, D.C.: The American Enterprise Institute Press, 2003.

Krimsky, Sheldon, and Dominic Golding. *Social Theories of Risk.* Westport, Conn.: Praeger, 1992.

Kuhn, Thomas S. *The Structure of Scientific Revolutions.* Chicago, University of Chicago Press, 1962.

Leonard, David. "NASA's Nuclear Prometheus Project Viewed as Major Paradigm Shift." http://www.space.com/business

technology//nuclear_power_030117.html (accessed 26 Mar 2004).

NASA. http://spacelink.nasa.gov/Instructional.Materials/ NASA.Educational.Products/International.Space.Station. Solar.Arrays/ISS.Solar.Arrays.pdf (accessed 5 Mar 2004).

NASA. Jet Propulsion Laboratory. http://www.jpl.nasa.gov/jimo/ (accessed 5 Mar 2004).

NASA. "Project Prometheus, Frequently Asked Questions, December 2003." http://www.jpl.nasa.gov/jimo/Schiff_FAQ_03_pdf2.pdf (accessed 11 Mar 2004).

National Science Board. *Science and Engineering Indicators, 2002.* Chapter 7: "Science and Technology: Public Attitudes and Public Understanding." http://www.nsf.gov/sbe/srs/seind02/c7/c7h.htm (accessed 27 Jan 2004).

———. "Science and Engineering Indictors, 2002." Figure 7–12. http://www.nsf.gov/sbe/srs/seind02/c7/fig07-12.htm (accessed 11 Mar 2004).

NERVA. http://www.astronautix.com/project/nerva.htm (accessed 24 Mar 2004).

Nye, Joseph S.; Philip D. Zelikow; and David C. King, eds. *Why People Don't Trust Government.* Cambridge, Mass.: Harvard University Press, 1997.

Online Newshour Forum. "Risks vs. Returns: Is the Cassini Mission Safe? October 21, 1997." http://www.pbs.org/newshour/forum/october97/cassini.html (accessed 24 Mar 2004).

Simpson, John W. *Nuclear Power from Undersea to Outer Space.* La Grange Park, Ill.: American Nuclear Society, 1995.

Sterns, Patricia M., and Leslie I. Tennen. "Regulation of Space Activities and Trans-science: Public Perceptions and Policy Considerations." *Space Policy* 11, no. 3 (Aug 1995): 182–85.

US Air Force. Space Command. *Strategic Master Plan for FY 04 and Beyond.* http://strategic-master-plan-04-beyond.pdf (accessed 26 Mar 2004).

US Department of Defense. *Report of the Commission to Assess United States National Security Space Management and*

Organization. 2001. http://www.defenselink.mil/space intro.pdf (accessed 26 Mar 2004).

US Department of Defense. *Report of the Secretary of Defense to the President and Congress, 1998.* http://www.defenselink.mil/execsec/adr98/chap8.html#top (accessed 26 Mar 2004).

US House. Science Committee. *Witnesses Suggest Change of Course for NASA Human Space Flight Programs, Testimony by Mr. Michael D. Griffin.* 108th Cong., 1st sess., 16 Oct 2003.

US Senate. Committee on Veteran Affairs, HUD and Independent Agencies. *Hearing on FY 03 NASA.* 107th Cong., 2d sess., 27 Feb 2002.

———. Committee on Commerce, Science, and Transportation. *In-Space Propulsion Technologies, Testimony by Mr. James H. Crocker, Vice-President of Lockheed Martin.* 108th Cong., 1st sess., 3 June 2003.

US White House. *Fact Sheet: National Space Policy.* 19 Sept 1996. http://www.ostp.gov/NSTC/html/fs/fs-5.html (accessed 26 Mar 2004).

Weinberg, Alvin M. *Science and Trans-science.* Alexandria, Va.: Minerva, 1972.

Flying Reactors
The Political Feasibility of Nuclear Power in Space

Air University Press Team

Chief Editor
Emily Adams

Copy Editor
Tammi Long

Book Design and Cover Art
Daniel Armstrong

Illustrations
Susan Fair

Composition and Prepress Production
Ann Bailey

Print Preparation
Joan Hickey

Distribution
Diane Clark

www.ingramcontent.com/pod-product-compliance
Lightning Source LLC
Chambersburg PA
CBHW081238180526
45171CB00005B/467